Open Verification Methodology Cookbook

Mark Glasser

Open Verification Methodology Cookbook

 Springer

Mark Glasser
Mentor Graphics Corporation
8005 SW. Boeckman Road
Wilsonville, OR 97070
USA
mark_glasser@mentor.com

ISBN 978-1-4419-0967-1   e-ISBN 978-1-4419-0968-8
DOI 10.1007/978-1-4419-0968-8
Springer Dordrecht Heidelberg London New York

Library of Congress Control Number: 2009930147

Printed on acid-free paper

Springer is part of Springer Science+Business Media (www.springer.com)

To Janice

To all verification engineers, the unsung heros of the design world, who toil over their testbenches so systems will work, and who receive satisfaction from a job well done.

Preface

When I need to learn a new piece of software I invent a little problem for myself that is within the domain of the application and then set out to solve it using the new tool. When the software package is a word processor, I'll use it to write a paper or article I'm working on; when the software is a drawing tool, I'll use it to draw some block diagrams of my latest creation. In the course of solving the problem, I learn how to use the tool and gain a practical perspective on which features of the tool are useful and which are not.

When the new software is a programming environment or a new programming language, the problem is a little different. I can't just apply the new language or environment to an existing problem. Unless I'm already familiar with the language, I don't want to commit to using it in a new development project. On the other hand, I may have an inkling that it would be best to use the new language. Otherwise, why would I be interested in it in the first place? I need a small program to help me understand the fundamental features and get a feel for how the language works. The program must be small and succinct, something I can write quickly and debug easily. Yet, it must use interesting language features I wish to learn.

Brian Kernighan and Dennis Ritchie solved this problem for all of us when they wrote the famous "Hello World" program. In their classic book *The C Programming Language*, they started off with a program that is arguably the most trivial program you could write in C that still does something. The beauty of Hello World is in its combination of simplicity and completeness. The program quoted here in its entirety, is not only simple, it also contains all of the constructs of a complete C program.

```
#include <stdio.h>

main()
{
    printf("hello, world\n");
}
```

All those years ago, I typed the program into my text editor, ran cc and ld, and a few seconds later saw my green CRT screen flicker with:

```
hello, world
```

Getting that simple program working gave me confidence that C was something I could conquer. I haven't counted how much C/C++ code I've written since, but it's probably many hundreds of thousands of lines. I've written all manner of software, from mundane database programs to exotic multi-threaded programs. It all started with Hello World.

The Open Verification Methodology (OVM) is a programming environment built upon SystemVerilog. It is designed to enable the development of complex testbenches. Like C (or SystemVerilog or SystemC), it will take some time and effort to study the OVM and understand how to apply all the concepts effectively. The goal of this book is to give you the confidence that running Hello World gave me all those years ago. If I, the author of this book, have done my job reasonably well, then somewhere along the way, as you read this book and exercise the examples, you should experience an aha! The metaphorical light bulb in your brain will turn on, and you will grasp the overall structure of the OVM and see how to apply it.

The premise of this book is that most engineers, like me, want to jump right into a new technology. They want to put their hands on it, try it out and see how it feels, learn the boundaries of what kinds of problems it addresses, and develop some practical experience. This is why quickstart guides and online help systems are popular. Generally, we do not want to read a lengthy manual and study the theory of operation first. We would rather plunge in, and later, refer to the manual only when and if we get stuck. In the meantime, as we experiment, we develop a general understanding of what the technology is and how to perform basic operations. Later, when we do crack open the manual, the details become much more meaningful.

This book takes a practical approach to learning about testbench construction. It provides a series of examples, each of which solves a particular verification problem. The examples are thoroughly documented and complete and delivered with build and run scripts that allow you to execute them in a simulator and observe their behavior. The examples are small and focused so you don't have to wade through a lot of ancillary material to get to the heart of an example.

This book presents the examples in a linear progression—from the most basic testbench, with just a pin-level stimulus generator, monitor, and DUT, to fairly sophisticated uses that involve stacked protocols, coverage, and automated testbench control. Each example in the progression introduces new concepts and shows you how to implement those concepts in a straightforward manner. Start by examining the first example. When you feel comfortable with it, move on to the second one. Continue in this manner, mastering each example and moving to the next.

The examples in the cookbook are there for you to explore. After you run an example, study the code to really understand its construction. The documentation provided with each example serves as a guidepost to point you to the salient features. Use this as a starting point to study the code organization, style, and other implementation details not explicitly discussed.

Play with the examples, too. Change the total time of simulation to see more results, modify the stimulus, add or remove components, insert print statements, and so on. Each new thing you try will help you more fully understand the examples and how they operate.

Feel free to use any of the code examples as templates for your work. For pieces that you find useful, cut and paste them into your code, or use them as a way to start developing your own verification infrastructures. Mainly, enjoy!

Mark Glasser, January 2009

Organization of This Book

Chapter 1. Describes some general principles of verification and establishes a framework for designing testbenches based on two questions—Does it work? and Are we done?

Chapter 2. This chapter provides an introduction to object-oriented programming and how OO techniques are applied to functional verification.

Chapter 3. Here, I introduce transaction-level modeling (TLM). The foundation of OVM is based on TLM. I illustrate basic put, get, and transport interfaces with examples.

Chapter 4. This chapter explains the mechanics of OVM, illustrating how to build hierarchies of class-based verification components and connect them with transaction-level interfaces. It also explains the essentials of using the OVM reporting facility.

Chapter 5. This chapter introduces the essential components of testbenches, such as drivers and monitors, and illustrates their construction with examples.

Chapter 6. This chapter discusses the essential topic of reuse—how to build components so that you have to do so only once and can apply what you have built in multiple situations.

Chapter 7. This chapter presents complete testbenches that use the types of components discussed so far and new ones, such as coverage collectors and scoreboards.

Chapter 8. OVM provides a facility called sequences for building complex stimulus generators. Sequences are discussed in this chapter, including how to construct sequences and how to use them to form a test API.

Chapter 9. It is important to reuse block-level testbenches when testing subassemblies or complete systems. This chapter illustrates some techniques for taking advantage of existing testbench components when constructing a system from separate blocks.

Chapter 10. SystemVerilog and OVM motivate new coding conventions. This chapter discusses some ways of constructing code to ensure that it is efficient, readable, and of course, reusable.

Obtaining the OVM Kit

You can get the open source OVM kit from www.ovmworld.com. The OVM kit contains complete source code and documentation.

Obtaining the Example Kit

The code used to illustrate concepts in this text is derived from the OVM cookbook kit available from Mentor Graphics. You can download the kit from *www.mentor.com*. Many of the snippets throughout the text have line numbers associated with them and, in some cases, a file name. The file names and line numbers are from the files in the Mentor OVM example kit.

Using the OVM Libraries

The OVM SystemVerilog libraries are encapsulated in a package called ovm_pkg. To use the package, you must import it into any file that uses any of the OVM facilities. The OVM library also contains a collection of macros that are useful in some places. You will need to include those as well as import the package

```
import ovm_pkig::*;
'include "ovm_macros.svh"
```

To make the OVM libraries available to your SystemVerilog testbench code, you must compile it into the work library. This requires two command line options when you compile your testbench with Verilog:

```
+incdir+<location-of-OVM-libraries>/src
<location-of-OVM-libraries>/src/ovm_pkg.sv
```

The first option directs the compiler to search the OVM source directory for include files. The second option identifies the OVM package to be compiled.

Building and Running the Examples

Installing the cookbook kit is a matter of unpacking the kit in a convenient location. No additional installation scripts or processes are required. You will have to set the OVM_HOME environment variable to point to your installation of OVM:

```
% setenv OVM_HOME <ovm-location>
```

Each example directory contains a run_questa script and one or more compile_* scripts. The run_questa script runs the example in its entirety. The compile script is a file that is supplied as an argument to the -f option on the compiler command line. Each example is also supplied with a vsim.do file that contains the simulator commands needed to run each example.

The simplest way to run an example is to execute its run_questa script:

```
% ./run_questa
```

This script compiles, links, and runs the example. You can also run the steps manually with the following series of commands:

```
% vlib work
% vlog -f compile_sv.f
% vsim -c top -do vsim.do
```

You must have the proper simulator license available to run the examples.

Who Should Read This Book?

This book is intended for electronic design engineers and verification engineers who are looking for ways to improve their efficiency and productivity in building testbenches and completing the verification portion of their projects. A familiarity with hardware description languages (HDL) in general, and specifically SystemVerilog, is assumed. It is also assumed that you know how to write programs in SystemVerilog, but it is not necessary to be an expert. Familiarity with object-oriented programming or OO terminology is helpful to fully understand the OVM. If you are not yet familiar with OO terminology, not to worry, the book introduces you to the fundamental concepts and terms.

Acknowledgements

The author wishes to acknowledge the people who contributed their time, expertise, wisdom, and in some cases, material to this project. This book would never have come to completion without their dedication to this project.

Without Adam Rose's simple, yet brilliant observation that construction of hierarchies of class-based components in SystemVerilog can be done in the same manner as SystemC, OVM and its predecessor AVM would not exist. Adam was also a key participant in the development of the TLM-1.0 standard which has greatly influenced the nature of OVM. Tom Fitzpatrick, who has been involved in the project since the earliest days of AVM, provided some material and helped refine the text. Rich Edelman, with humor, good grace, and a keen eye for detail, and Andy Meyer, with his amazingly deep reservoir of verification knowledge, allowed me to bounce ideas around with them and helped me crystallize the concepts and flow of the material. Adam Erickson, who is a true code wizard and an expert in object-oriented patterns, always keeps me honest.

Todd Burkholder taught me about narrative flow and loaned me some if his English language skills to smooth out awkward sentences. Jeanne Foster did the detailed copy editing and an insightful job of producing an index.

Thanks to Harry Foster, who inspired the HFPB protocol and encouraged me to write this book. Hans VanderSchoot did a detailed review of the text and suggested many good ideas for improving the text. Also thanks to Kurt Schwartz of WHDL who reviewed an early draft. Cliff Cummings provided excellent advice on construction of the RTL examples.

A special thanks to Jan Johnson, who sponsored and supported the OVM Cookbook project from its inception.

Contents

List of Figures

Introduction

Software construction is not usually a topic that immediately comes to mind when hardware designers and verification engineers talk about their work. Designers and verification engineers, particularly those schooled in electrical engineering, naturally think of design and verification work as a "hardware problem," meaning that principles of hardware design are required to build and verify systems. Of course, they are largely, but not entirely, correct. Electronic design requires an in-depth knowledge of hardware, everything from basic DC and AC circuit analysis and transistor operation to communication protocols and computer architecture. A smattering of physics is useful too for designs implemented in silicon (which is the intent for most). However, building a testbench to verify a hardware design is a different kind of problem—it is a software problem.

Today, with the availability of reliable synthesizers and the application of synchronous design techniques, the lowest level of detail that designers must consider is register transfer level (RTL). As the name suggests, the primary elements of a design represented at this level are registers, interconnections between registers, and the computation necessary to modify their values. Since each register receives new values only when the clock pulses, all of the combinational logic needed to compute the register value can be abstracted to a set of Boolean and algebraic expressions.

RTL straddles the hardware and software worlds. The components of an RTL design are readily identifiable as hardware; such as registers, wires, and clocks. Yet the combinational expressions and control logic look suspiciously like those in typical procedural programming languages, such as C or Java. The process of building an RTL design is much like programming. You write code that represents the structures in your design using an HDL, a special programming language designed specifically for this purpose. You use compilers, linkers, and debuggers, just as you would if you were programming in C. There are differences, of course. You do not need to consider issues surrounding timing, concurrency, and synchronization when programming in C (unless you are writing embedded software, which further blurs the line between hardware and software).

Testbenches live squarely in the software world. The elements of a testbench are exactly the same as those found in any software system—data structures and algorithms. Testbenches are *hardware aware* since their job is to control, respond to, and analyze hardware. Still, the bulk of their construction and operation falls under the software umbrella.

Most of what a testbench "does," does not involve hardware. Testbenches operate at levels of abstraction higher than RTL, thus they do not require registers, wires, and other hardware elements. We can categorize the testbench results we collect and analyze as data processing, which does not involve hardware elements at all. Testbench programs do not need to be implemented in silicon, which completely frees them from the limitations of synthesizable constructs. The only place that a testbench is involved with hardware is at its interfaces. Testbenches must stimulate and respond to hardware. Testbenches must *know about* hardware, but they do not need to *be* hardware.

Because testbenches are software, it is appropriate to apply software construction techniques to building them. Software construction is at the center of modern verification technology and the OVM. Software construction is itself a very large topic on which many volumes have been written. It is not possible for us to go into great depth on topics such as object-oriented programming, library organization, code refactoring, testing strategies, and so on. However, this book touches on these topics in a practical way, showing how to apply software techniques to building testbenches. I rely heavily on examples to illustrate the principles discussed.

1

Verification Principles

This chapter surveys general principles of verification and establishes a framework for designing testbenches based on two questions—Does it work? and Are we done?

1.1 Verification Basics

Functionally verifying a design means comparing the designer's intent with observed behavior to determine their equivalence. We consider a design verified when, to everyone's satisfaction, it performs according to the designer's intent. This basic principle often gets lost in the discussion of testbenches, assertions, debuggers, simulators, and all the other paraphernalia used in modern verification flows. To tell if a design works, you must compare it with some known reference that represents the designer's intent. Keep this in mind as you read the rest of this book. Every testbench has some kind of reference model and a means to compare the function of the design with the reference.

When we say "design," we mean the design being verified, often called the design under test or DUT. To be verified, the DUT is typically in some form suitable for production—a representation that can be transformed into silicon by a combination of automated and manual means. We distinguish a DUT from a sketch on the back of a napkin or a final packaged die, neither of which is in a form that can be verified. A reference design captures the designer's *intent*, that is, what the designer expects the design to do. The reference can take many forms, such as a document describing the operation of the DUT, a golden model that contains a unique algorithm, or assertions that represent a protocol.

M. Glasser, *Open Verification Methodology Cookbook*, DOI: 10.1007/978-1-4419-0968-8_1,
© Mentor Graphics Corporation, 2009

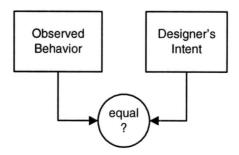

Figure 1-1 Comparing Design and Intent

To automate the comparison of the behavior with the intent, both must be in a form that we can execute on a computer using some program that does the comparison. Exactly how to do this is the focus of the rest of this book. The problem of building a design is a topic beyond the scope of this text. Here, we confine our discussion to the problem of capturing design intent and comparing it with the design to show their equivalence.

1.1.1   Two Questions

Verifying a design involves answering two questions: Does it work? and Are we done? These are basic, and some would say, obvious questions. Yet they motivate all the mechanics of every verification flow. The first question is Does it work? This question comes from the essential idea of verification we discussed in the previous section. It asks, Does the design match the intent? The second question is Are we done? It asks if we have satisfactorily compared the design and intent to conclude whether the design does indeed match the intent, or if not, why not. We use these valuable questions to create a framework for developing effective testbenches.

1.1.2   Does It Work?

Does it work? is not a single question but a category of questions that represent the nature of the DUT. Each design will have its own set of does-it-work questions whose role is to determine *functional correctness* of the design. Functional correctness questions ask whether the device behaves properly in specific situations. We derive these questions directly from the design intent, and we use them to express design intent in a testbench.

Consider a simple packet router as an example. This device routes packets from an input to one of four output ports. Packets contain the address of the destination port and have varying length payloads. Each packet has a header, trailer, and two bytes for the cyclic redundancy check (CRC). The does-it-work questions might include these:

- Does a packet entering the input port addressed to output port 3 arrive properly at port 3?
- Does a packet of length 16 arrive intact?
- Are the CRC bytes correct when the payload is [0f 73 a8 c2 3e 57 11 0a 88 ff 00 00 33 2b 4c 89]?

This is just a sample of a complete set of questions. For a device even as relatively simple as this hypothetical packet router, the set of does-it-work questions can be long. To build a verification plan and a testbench that supports the plan, you must first enumerate all the questions or show how to generate all of them, and then select the ones that are interesting.

Continuing with the packet router example, to enumerate all the does-it-work questions, you can create a chart like this:

Number	Does-It-Work Questions
1.	For all four output ports, does a packet arriving at the input port addressed to an output port arrive at the proper output port?
2.	Do packets of varying payload sizes, from eight bytes to 256 bytes, arrive intact?
3.	Is the CRC computation correct for every packet?
4.	Is a packet with an incorrect header flagged as an error?

The table above contains two kinds of questions, those we can answer directly and those we can break down into more detailed questions. Question 1 is a series of questions we can explicitly enumerate:

Number	Does-It-Work Questions
1a	Does a packet arriving at the input port addressed to output port 0 arrive at port 0?
1b	Does a packet arriving at the input port addressed to output port 1 arrive at port 1?

Number	Does-It-Work Questions
1c	Does a packet arriving at the input port addressed to output port 2 arrive at port 2?
1d	Does a packet arriving at the input port addressed to output port 3 arrive at port 3?

Notice that we formulate all of the questions so that they can be answered yes or no. At the end of the day, a design either works or it doesn't—it either is ready for synthesis and place and route or it is not. If you can answer all the questions affirmatively, then you know the design is ready for the next production step.

When you design your set of does-it-work questions, remember to word them so they can be answered yes or no. A yes answer is positive; that is, answering yes means the device operates correctly. That will make things easier than trying to keep track of which questions should be answered yes and which should be answered no. A question such as Did the router pass any bad packets? requires a no answer to be considered successful. A better wording of the question is, Did the router reject bad packets? But you should make the questions as specific as you can, so an even better wording is, When a bad packet entered the input port, did the router detect it, raise the error signal, and drop the packet? Keep in mind that more specific questions tell you more about the machinery. Your testbench needs to determine the yes or no answer.

A properly worded yes or no question contains its own success criteria. It says what will achieve a yes response. A question such as, Is the average latency less than 27 clock cycles? contains the metric, 27 clock cycles, and the form of comparison, less than. If the question is (improperly) worded as, What is the average latency of packets through the router? we will not know what is considered acceptable. To answer either question, you first must be able to determine the average latency. Only in the correct wording of the question do we know how to make a comparison to determine whether the result is correct. The metric by itself does not tell us whether the design is functioning as intended. When we compare the measured value against the specification, 27 clock cycles in this example, we can determine whether the design works.

As is often the case, it is not practical to enumerate every single does-it-work question. To verify that every word in a 1 Mb memory can be written to and read from, it is neither practical nor necessary to write one million questions. Instead, a *generator question*, a question that generates many others, takes the place of one million individual questions. Can each of the one million words in the memory be successfully written to and read from? is a generator question.

Other questions may themselves represent classes of questions. Question 3, Is the CRC computation correct for every packet? is an example. Testing the CRC computation requires a number of carefully-thought-through tests to determine whether the CRC computation is correct in all cases. For example, we also want to test what happens when the payload is all zeros, is all ones, has an odd number of bytes, has an even number of bytes, has odd bytes that are all zero and even bytes that are all one, and so forth.

1.1.3   Are We Done?

To determine the answer to Are we done?, we need to know if we have answered enough of the does-it-work questions to claim that we have sufficiently verified the design. We begin this task by prioritizing all the does-it-work questions across two axes:

	Easy to answer	Hard to answer
Most critical functionality	No-brainer.	Get to work!
Least critical functionality	Probably can omit.	Don't waste the time.

The art of building a testbench requires that, in the initial stage, we identify the set of questions and sort them to identify the ones that return the highest value in terms of verifying the design. The next step is to build the machinery that will answer the questions and determine which ones have been answered (and which have not).

Are-we-done questions are also called *functional coverage* questions, questions that relate to whether the design is sufficiently covered by the test suite in terms of design function. As with does-it-work questions, we can also decompose functional coverage questions into more detailed questions. And just like functional correctness questions, functional coverage questions must also be answerable in terms of yes or no. The following list includes examples of functional coverage questions:

- Has every processor instruction been executed at least once?
- Has at least one packet traversed from every input port to every output port?

- ■ Has the memory been successfully accessed with a set of
  addresses that exercise each address bit as one and then each
  address bit as zero, not including `0xffffffff` and `0x00000000`?

Another way to think of these questions is that they ask, Have the necessary
does-it-work questions been answered affirmatively? When we think of
functional coverage in this light, the term refers to covering the set of does-it-
work questions. Furthermore, coverage questions identify a metric and a
threshold for comparison. Coverage is reached (that is, the coverage question
can be answered yes) when the metric reaches the threshold.

In summary, the art of building a testbench begins with a test plan. The test
plan begins with a carefully thought out set of does-it-work and are-we-done
questions.

1.1.4   Two-Loop Flow

The process for answering the does-it-work and are-we-done questions can
be described in a simple flow diagram as shown in Figure 1-2. Everything is
driven by the functional specification for the design. From the functional
specification, you can derive the design itself and the verification plan. The
verification plan drives the testbench construction.

The flow contains two loops, the does-it-work loop and the are-we-done loop.
Both loops start with a simulation operation. The simulation exercises the
design with the testbench and generates information we use to answer the
questions. First we ask, Does it work? If any answer is no, then we must
debug the design. This debugging exercise can result in changes to the design
implementation.

Once the design works to the extent it has been tested, then it is time to
answer the question Are we done? We answer this question by collecting
coverage information and comparing it against thresholds specified in the test
plan. If we do not reach those thresholds, then the answer is no, and we must
modify the testbench to increase the coverage. Then we simulate again.

Changing the testbench or the stimulus can cause other latent design bugs to
surface. A subsequent iteration around the loop may cause us to go back to
the does-it-work loop again to fix any new bugs that appear. As you can see, a
complete verification process flip-flops back and forth between does-it-work

and are-we-done loops until we can answer yes for all the questions in both categories.

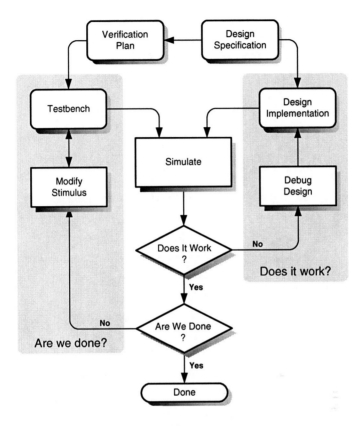

Figure 1-2 Two-Loop Flow

In an ideal world, a design has no bugs and the coverage is always sufficient, so you only have to go around each loop once to get yes answers to both questions. In the real world, it can take many iterations to achieve two yes answers. One objective of a good verification flow is to minimize the number of iterations to complete the verification project in the shortest amount of time using the smallest number of resources.

1.2   First Testbench

Let's jump right in by illustrating how to verify one of the most fundamental devices in a digital electronic design, an AND gate. An AND gate computes the logical *and* of the inputs. The function of this device is trivial, and in practice, is hardly worth its own testbench. Because it is trivial, we can use it to illustrate some basic principles of verification without having to delve into the details of a more complex design.

Figure 1-3 shows the schematic symbol for a two-input AND gate. The gate has two inputs, A and B, and a single output Y. The device computes the logical AND of A and B and puts the result on Y.

Figure 1-3 A Two-Input AND Gate

The following truth table describes the function of the device.

A	B	Y
0	0	0
0	1	0
1	0	0
1	1	1

The truth table is exhaustive: it contains all possible inputs for A and B and thus all possible correct values for output Y.

Our mission is to prove that our design, the AND gate, works correctly. To verify that it does indeed perform the AND function correctly, we first need to list the questions. The truth table helps us create the set of questions we need to verify the design. Each row of the table contains an input for A and B and the expected output for Y. Since the table is exhaustive, our generator question is, For each row in the truth table, when we apply the values of A and B identified in that row, does the device produce the expected output for Y? To answer the are-we-done question, we determine whether we have exercised each row in the truth table and received a yes answer to the does-it-work question for that row. Our are-we-done question is Do all rows work?

To automate answering both the does-it-work and are-we-done questions, we need some paraphernalia, including the following:

- A model that represents the DUT (in this case, the AND gate)
- The design intent in a form we can codify as a reference model

- Some stimuli to exercise the design
- A way to compare the result to the design intent

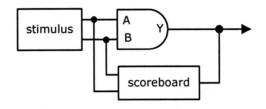

Figure 1-4 First Testbench

While our little testbench is simple, it contains key elements found in most testbenches at any level of complexity. The key elements are the following:

- DUT
- Stimulus generator — generates a sequence of stimuli for the DUT
- Scoreboard — embodies the reference model

The scoreboard observes the inputs and outputs of the DUT, performs the same function as the DUT except at a higher level of abstraction, and determines whether the DUT and reference match. The scoreboard helps us answer the does-it-work questions.

1.2.1 DUT

The DUT is our two-input AND gate. We implement the AND gate as a module with two inputs, A and B, and one output Y.

```
44      module and2 (
45         output bit Y,
46         input A, B);
47
48         initial Y = A & B;
49
50         always @* Y = #1 A & B;
51      endmodule
```

The stimulus generator in this example generates *directed stimulus*. Each new value emitted is computed in a specific order. Later, we will look at *random stimulus generators* which, as their name suggests, generate random values.

```
81     module stimulus(output bit A, B);
82
83        bit [1:0] stimulus_count = 0;
84
85        always
86           #10 {A,B} = stimulus_count++;
87
88     endmodule
```

The purpose of the stimulus generator is to produce values as inputs to the DUT. stimulus, a two-bit quantity, contains the value to be assigned to A and B. After it is incremented in each successive iteration, the low-order bit is assigned to A, and the high-order bit is assigned to B.

1.2.2 Scoreboard

The *scoreboard* is responsible for answering the does-it-work question. It watches the activity on the DUT and reports whether it operated correctly.[1] One important thing to notice is that the structure of the scoreboard is strikingly similar to the structure of the DUT. This makes sense when you consider that the purpose of the scoreboard is to track the activity of the DUT and determine whether the DUT is working as expected.

```
59     module scoreboard(input bit Y, A, B);
60
61        reg Y_sb, truth_table[2][2];
62
63        initial begin
64           truth_table[0][0] = 0;
65           truth_table[0][1] = 0;
66           truth_table[1][0] = 0;
67           truth_table[1][1] = 1;
68        end
69
70        always @(A or B) begin
71           Y_sb = truth_table[A][B];
72           #2 $display("@%4t - %b%b : Y_sb=%b, Y=%b (%0s)",
73                    $time, A, B, Y_sb, Y,
74                    ((Y_sb == Y) ? "Match" : "Mis-match"));
75        end
76     endmodule
```

---

1. For anything more sophisticated than an AND gate, the monitor and response checker would be separate components in the testbench. For the trivial AND gate testbench, this would be more trouble than it's worth and would cloud the basic principles being illustrated.

The scoreboard pins are all inputs. The scoreboard does not cause activity on the design. It passively watches the inputs and outputs of the DUT.

The top-level module, shown below, is completely structural; it contains instantiations of the DUT, the scoreboard, and the stimulus generator, along with the code necessary to connect them.

```
93      module top;
94          stimulus    s(A, B);
95          and2        a(Y, A, B);
96          scoreboard sb(Y, A, B);
97
98          initial
99              #100 $finish(2);
100     endmodule
```

When we run the simulation for a few iterations, here is what we get:

```
# @  22 - 01 : Y_sb=0, Y=0 (Match)
# @  32 - 10 : Y_sb=0, Y=0 (Match)
# @  42 - 11 : Y_sb=1, Y=1 (Match)
# @  52 - 00 : Y_sb=0, Y=0 (Match)
# @  62 - 01 : Y_sb=0, Y=0 (Match)
# @  72 - 10 : Y_sb=0, Y=0 (Match)
# @  82 - 11 : Y_sb=1, Y=1 (Match)
# @  92 - 00 : Y_sb=0, Y=0 (Match)
```

Each message has two parts. The first part shows the stimulus being applied. The second part shows the result of the scoreboard check that compares the DUT's response to the predicted response. We use a colon to separate the two parts.

This simple testbench illustrates the use of a stimulus generator and a scoreboard that serves as a reference. Although the DUT is a simple AND gate, all the elements of a complete testbench are present.

1.3   Second Testbench

The previous example illustrated elementary verification concepts using a combinational design, an AND gate. Combinational designs, by their very nature, do not maintain any state data. In our second example, we look at a slightly more complex design that maintains state data and uses a clock to cause transitions between states.

The verification problem associated with synchronous (sequential) designs is a little different than for combinational designs. Everything you need to know about a combinational design is available at its pins. A reference model for a combinational device simply needs to compute $f(x)$ where x represents the inputs to the device and f is the function it implements. The outputs of a sequential device are a function of its inputs and its internal state. Further computation may change the internal state. The scoreboard must track the internal state of the DUT and compare the output pins.

A combinational device can be exhaustively verified by exercising all possible inputs. For a device with n input pins, we must apply a total of 2^n input vectors. The number 2^n can be large, but computing that many inputs is easy. We just need to have an n-bit counter and apply each value of the counter to the inputs of the device.

For a sequential device, the notion of "done" must extend to covering not only the total number of possible inputs, but also the number of possible internal states. For a device with n inputs and m internal states, you must cover (2^n inputs) * (2^m states), which is 2^{n+m} combinations of internal states and inputs. For a device with 64 input pins and a single 32-bit internal register, the number of state-input combinations is 2^{96} — a very large number indeed!

Even for very large numbers of combinations, the verification problem would not be too difficult if it were possible to simply increment a counter to reach all combinations, as we do with combinational devices. Unfortunately, that is not possible. The internal state is not directly accessible from outside the device. It can only be modified by manipulating the inputs. The problem now becomes how to reach all the states in the device through only manipulating the inputs. This is a difficult problem that requires a deep understanding of the device to generate sequences of inputs to reach all the states.

Since it is difficult to reach all the states, the obvious question becomes, Can we prune the problem by reducing the number of states that we need to reach to show that the device works correctly? The answer is yes. Now the question becomes, How do we decide which states do not need to be covered?

This topic is complex, and a full treatment of it is beyond the scope of this text. However, we can give an intuitive answer to the question. States that can be shown to be unreachable, through formal verification or other techniques, do not need to be covered. The designer should consider simplifying the design to remove unreachable states, since they provide no value. States that have a low probability of being reached may also be eliminated from the verification plan. Determining the probability threshold and assigning probabilities to states is as much an art as a science. It involves understanding how the design is used and which inputs are expected (as compared to which are possible).

It is also possible to eliminate coverage of states that are functionally equivalent. Consider a packet communications device. In theory, every possible packet payload value represents a distinct state (or set of states) as it passes through the design, and it should be covered during verification. However, it is probably not a stretch to consider that arbitrary non-zero values are, for all intents and purposes, equivalent. Of course, there might be some interesting corner cases that must be checked, such as all zeros, all ones, particular values that might challenge the error correction algorithms, and so forth. Variations in data become interesting when they affect control flow.

In general, it is more important to cover control states than data states. A common way to reduce the number of states necessary to cover a design is to separate data and control. For a particular control path, the data can be arbitrary. For certain data, you may want to fix the control path. For example, in an ALU, a design that we will consider in detail in later chapters, you can separate the control functions of getting data into and out of the registers and establishing the arithmetic operation to be performed from the results of specific arithmetic operations. Using directed control, you can randomize data or look at data corner cases such as divide by 0 or multiply by 1.

For complex sequential designs, determining which states to cover (and which do not need to be covered) and how to reach those states with minimal effort is a problem that keeps verification engineers employed. In this section, we will consider a small sequential device whose internal states can easily be covered.

1.3.1 3-Bit Counter

The design shown in Figure 1-5 is a 3-bit counter with an asynchronous reset. Each time the clock pulses high, the count increments. The design is composed of three toggle flip-flops, each of which maintains a single bit of the counter. The flip-flops are connected with some combinational logic to form a counter. Each flip-flop toggles when the T input is high. When T is low, the flip-flop maintains its current state. When the active low reset is set to 0, the flip-flop moves to a 0 state.

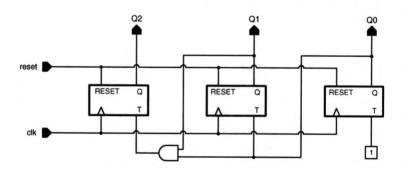

Figure 1-5 3-Bit Counter

The code for the counter is contained in two modules. One is a simple toggle flip-flop, and the other connects the flip-flops with the necessary glue logic to form a counter. The first example shown below is the toggle flip-flop.

```
36     module toggle_ff (output bit q, input t, rst_n, clk);
37
38        always @ (posedge clk or negedge rst_n)
39           if (!rst_n) q <= '0;
40           else if (t) q <= ~q;
41
42     endmodule
```

The counter comprises three toggle flip-flops and an AND gate.

```
47     module counter (output [2:0] q, input rst_n, clk);
48
49        toggle_ff ff0 (q[0], 1'b1, rst_n, clk);
50        toggle_ff ff1 (q[1], q[0], rst_n, clk);
51        toggle_ff ff2 (q[2], t2,   rst_n, clk);
52        and a1 (t2, q[0], q[1]);
53
54     endmodule
```

The design is straightforward, but it has characteristics that are common in real designs and that require some attention for proper design verification. The key characteristics are that the design is driven by a clock, and that it maintains internal state. The AND gate from the previous example does not maintain any state. All of the information about what the design is doing can be gleaned from its pins. In a design with internal data, that is not the case.

This difference is reflected in the design of our scoreboard. Figure 1-6 shows the organization for the testbench for the 3-bit counter.

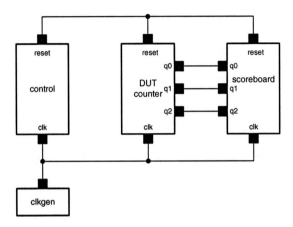

Figure 1-6 Testbench Organization for 3-Bit Counter

In many respects, the testbench for the 3-bit counter is much like the one for the AND gate. Both have a scoreboard whose role is to watch what the design is doing and determine whether it is working correctly. Both have a device for driving the DUT. However, we manage operation differently for these designs. We use a stimulus generator for the AND gate, but we use a *controller* for the 3-bit counter. The 3-bit counter is a free-running device. As long as it is connected to a running clock, it will continue to count. So we do not need a stimulus generator as we did with the AND gate. Instead, the controller manages the operation of the DUT and testbench. The controller provides an initial reset so that the count starts from a known value. It also stops the simulation at the appropriate time.

```
93      module control(output bit rst_n, input clk);
94
95        initial begin
96          rst_n <= 0;
97          @(posedge clk);
98          @(negedge clk);
99          rst_n <= 1;
100         repeat (10) @(posedge clk);
101         $finish;
102       end
103
104     endmodule
```

The scoreboard must track the internal state of the DUT. It does this using the variable count. Like the DUT, when reset is activated, count is set to 0. Each clock cycle count increments, and the new value is compared with the count from the DUT.

```
71     module scoreboard (input [2:0] q, input rst_n, clk );
72
73       int count;
74
75       always @(posedge clk or negedge rst_n) begin
76         if(!rst_n) count <= 0;
77         else begin
78           if (count == q)
79             $display("time =%4t q = %3b count = %0d match!",
80                       $time, q, count);
81           else
82             $display("time =%4t q = %3b count = %0d <-- no
match",
83                       $time, q, count);
84           count <= (count + 1) % 8;
85         end
86       end
87
88     endmodule
```

The scoreboard has a high-level model of the counter. It uses an integer variable and the plus (+) operator to form a counter instead of flip-flops and AND gates. Each time the clock pulses, it increments its count, just like the RTL counter. It also compares to see if its internal count matches the output of the counter DUT.

For completeness, the example shows the clock generator and top-level module. The clock generator simply initializes the clock to zero, and then it toggles it every 5 ns.

```
59     module clkgen(output bit clk);
60
61       initial  begin
62         clk <= 0;
63         forever #5 clk = ~clk;
64       end
65
66     endmodule
```

The top-level module is typical of most testbenches. It connects the DUT and the testbench components.

```
109    module top;
110
111        wire [2:0] q;
112
113        clkgen   ckgn      (clk);
114        counter  cntr      (q, rst_n, clk);
115        control  ctrl      (rst_n, clk);
116        scoreboard score   (q, rst_n, clk);
117
118    endmodule
```

We have illustrated a simple testbench that contains the elements used in much more sophisticated testbenches. Sequential designs that maintain internal state require a scoreboard that operates in parallel with the DUT. The scoreboard performs the same computations as the DUT but at a higher level of abstraction. The scoreboard also compares its own computation with inputs received from the DUT.

1.4   Layered Organization of Testbenches

Just as a design is a network of design components, a testbench is a network of verification components. The OVM defines verification components, their structure, and interfaces. This section describes the essential OVM components.

OVM testbenches are organized in layers. The bottommost layer is the DUT, an RTL device with pin-level interfaces. Above that is a layer of transactors, devices that convert between the transaction-level and pin-level worlds. The components in the layers above the transactor layer are all transaction-level components. The diagram below illustrates the layered testbench organization. The box on the left identifies the name of the layer. The box on the right lists the type of components in that layer. The vertical arrows show which layers communicate directly. For example, the control layer communicates with the analysis, operational, and transactor layers, but not directly with the DUT.

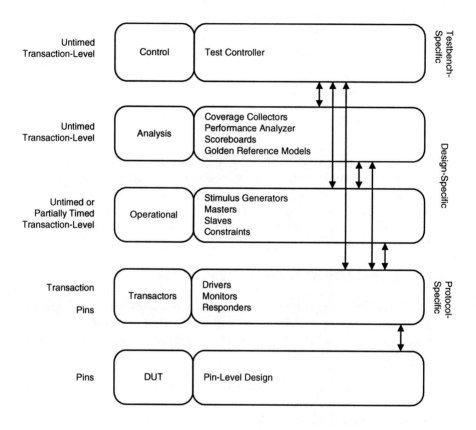

Figure 1-7 OVM Testbench Architecture Layers

You can also view a testbench as a concentric organization of components. The innermost ring maps to the bottom layer, and the outermost ring maps to the top layer. Some find it easier to understand the relationships between the layers using a netlist style diagram.

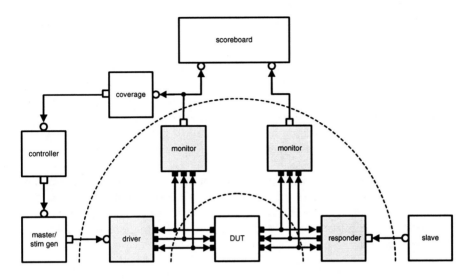

Figure 1-8 Concentric Testbench Organization

1.4.1   Transactors

The role of a transactor in a testbench is to convert a stream of transactions to pin-level activity or vice versa. Transactors are characterized by having at least one pin-level interface and at least one transaction-level interface. Transactors come in a wide variety of shapes, colors, and styles. We'll focus on monitors, drivers, and responders.

Monitor. A *monitor*, as the name implies, monitors a bus. It watches the pins and converts their wiggles to a stream of transactions. Monitors are passive, meaning they do not affect the operation of the DUT in any way.

Driver. A *driver* converts a stream of transactions (or sequence items) into pin-level activity.

Responder. A *responder* is much like a driver, but it responds to activity on pins rather than initiating activity.

1.4.2   Operational Components

The *operational components* are the set of components that provide all the things the DUT needs to operate. The operational components are responsible for generating traffic for the DUT. They are all transaction-level components and have only transaction-level interfaces. The ways to generate stimulus are as varied as the kinds of devices there are to verify. We'll look at three general kinds of operational components: stimulus generators, masters, and slaves.

Stimulus Generator. *Stimulus generators* create a stream of transactions for exercising the DUT. Stimulus generators can be random, directed, or directed random; they can be free running or have controls; and they can be independent or synchronized. The simplest stimulus generator randomizes the contents of a request object and sends that object to a driver. OVM also provides a modular, dynamic facility for building complex stimulus called sequences. These are discussed in detail in Chapter 8.

Master. A *master* is a bidirectional component that sends requests and receives responses. Masters initiate activity. Like stimulus generators, they can send individual randomized transactions or sequences of directed or directed-random transactions. Masters may use the responses to determine their next course of action. Masters can also be implemented in terms of sequences.

Slave. *Slaves*, like masters, are bidirectional components. They respond to requests and return responses (in contrast to masters, which send requests and receive responses).

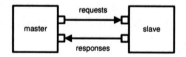

Figure 1-9 A Master and a Slave

1.4.3   Analysis Components

Analysis components receive information about what's going on in the testbench and use that information to make some determination about the correctness or completeness of the test. Two common kinds of analysis components are scoreboards and coverage collectors.

Scoreboard. *Scoreboards* are used to determine correctness of the DUT, to answer does-it-work questions. Scoreboards tap off information going into and out of the DUT and determine if the DUT is responding correctly to its stimulus.

Coverage Collector. *Coverage collectors* count things. They tap into streams of transactions and count the transactions or various aspects of the transactions. The purpose is to determine verification completeness by answering are-we-done questions. The particular things that a coverage collector counts depends on the design and the specifics of the test. Common things that coverage collectors count include: raw transactions, transactions that occur in a particular segment of address space, and protocol errors. The list is limitless.

Coverage collectors can also perform computations as part of a completeness check. For example, a coverage collector might keep a running mean and standard deviation of data being tracked. Or it might keep a ratio of errors to good transactions.

1.4.4 Controller

Controllers form the main thread of a test and orchestrate the activity. Typically, controllers receive information from scoreboards and coverage collectors and send information to environment components.

For example, a controller might start a stimulus generator running and then wait for a signal from a coverage collector to notify it when the test is complete. The controller, in turn, stops the stimulus generator. More elaborate variations on this theme are possible. In an example of a possible configuration, a controller supplies a stimulus generator with an initial set of constraints and starts the stimulus generator running. When a particular ratio of packet types is achieved, the coverage collector signals the controller. Rather than stopping the stimulus generator, the controller may send it a new set of constraints.

1.5   Two Domains

We can view the set of components in a testbench as belonging to two separate domains. The *operational domain* is the set of components, including the DUT, that operate the DUT. These are the stimulus generators, bus functional models (BFM), and similar components that generate stimulus and provide responses that drive the simulation. The DUT, responder, and driver transactions—along with the environment components that directly feed or respond to drivers and responders—comprise the operational domain. The rest of the testbench components—monitor transactors, scoreboards, coverage collectors, and controller—comprise the *analysis domain*. These are the components that collect information from the operational domain.

Data must be moved from the operational domain to the analysis domain in a way that does not interfere with the operation of the DUT and preserves event timing. This is accomplished with a special communication interface called an *analysis port*. Analysis ports are a special kind of transaction port in which a publisher broadcasts data to one or more subscribers. The publisher signals all the subscribers when it has new data ready.

One of the key features of analysis ports is that they have a single interface function, write(). Analysis FIFOs, the channels used to connect analysis ports to analysis components, are unbounded. This guarantees that the publisher doesn't block and that it deposits its data into the analysis FIFO in precisely the same delta cycle in which it was generated. Chapter 7 discusses analysis ports and analysis FIFOs in more detail.

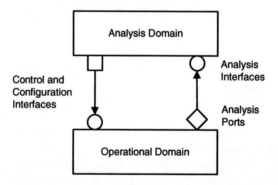

Figure 1-10   Connection between Operational and Analysis Domains

Generally, the operational and analysis domains are connected by analysis ports and control and configuration interfaces. Analysis ports tap off data concerning the operation of the DUT. These data might include bus transactions, communication packets, and status information (success or failure of specific operations). The components in the analysis domain analyze the data and make decisions. The results of those decisions can be communicated to the operational domain via the control and configuration interfaces. Control and configuration interfaces can be used to start and stop stimulus generators, change constraints, modify error rates, or manipulate other parameters affecting how the testbench operates.

1.6   Summary

In this chapter we looked at how to structure an overall verification process. The process is based on two fundamental questions, Does it work? and Are we done? Simple examples illustrated how to build testbench machinery to answer these questions with devices such as stimulus generators and scoreboards. The rest of this book shows how to apply transaction-level modeling techniques to build practical, scalable, reusable testbench components that answer these questions and shows how to connect them to form testbenches.

2

Fundamentals of Object-Oriented Programming

Software engineering, unconstrained by the physics of electricity and magnetism, has long sought to build reusable, interchangeable, robust components. An important programming model that addresses the problem is called *object-oriented programming* (OOP). The central idea of OOP is that programs are organized as a collection of interacting objects, each with its own data space and functions. Objects can be made reusable because they encapsulate everything they need to operate, can be built with minimal or no external dependencies, and can be highly parameterized.

This chapter introduces the basic concepts of OOP, including the notions of encapsulation and interface. The chapter concludes with a discussion of why OOP is important for building testbenches.

2.1   Procedural vs. OOP

To understand OOP and the role it plays in verification, it is beneficial to first understand traditional procedural programming and its limitations. This sets the foundation for understanding how OOP can overcome those limitations.

In the early days of assembly language programing, programmers and computer architects quickly discovered that programs often contained sequences of instructions that were repeated throughout a program. Repeating lots of code (particularly with a card punch) is tedious and error prone. Making a change to the sequence involved locating each place the sequence appeared in the program and repeating the change in each location. To avoid the tedium and the errors caused by repeated sequences, the subroutine was invented.

M. Glasser, *Open Verification Methodology Cookbook*, DOI: 10.1007/978-1-4419-0968-8_2,
© Mentor Graphics Corporation, 2009

A subroutine is a unit of reusable code. Instead of coding the same sequence of instructions inline, you call a subroutine. Parameters passed to subroutines allow you to dynamically modify the code. That is, each call to a subroutine with different values for the parameters causes the subroutine to behave differently based on the specific parameter values.

Every programming language of any significance has constructs for creating subroutines, procedures, or functions, along with syntax for passing in parameters and returning values. These features are useful for creating operations that are used often. However, some operations are very common (such as I/O, data conversions, numerical methods, and so forth). And to avoid having to rewrite these operations repeatedly, programmers found it valuable to create libraries of commonly used functions. As a result, most programming languages include such a library as part of the compiler package. One of the most well-known examples is the C library that comes with every C compiler. It contains useful functions such as `printf()`, `cos()`, `atof()`, and `qsort()`. These are functions that virtually every programmer will use at some time or another.

Imagine having to write your own I/O routines or your own computation for converting numbers to strings and strings to numbers. There was a time when programmers did just that. Libraries of reusable functions changed all that and increased overall programming productivity.

As software practice and technology advanced, programmers began thinking at higher levels of abstraction than instructions and subroutines. Instead of writing individual instructions, programmers now code in languages that provide highly abstracted models of the computer, and compilers or interpreters translate these models into specific instructions. A library, such as the C library or STL in C++, is a form of abstraction. It presents a set of functions that programmers can use to construct ever more complex programs or abstractions.

In his seminal book *Algorithms + Data Structures = Programs*, Niklaus Wirth explains that to solve any programming problem, you must devise an abstraction of reality that has the characteristics and properties of the problem at hand and ignore the rest of the details. He argues that the collection of data you need to solve a problem forms the abstraction. So before you can solve a problem, you first need to determine what data you need to have to create the solution.

To continue building reusable abstractions, we need to create libraries of data objects that can be reused to solve specific kinds of problems. The search for ways to do this leads to the development of object-oriented technology.

Object-oriented program analysis and design is centered around data objects, the functionality associated with each object, and the relationships between objects.

The goal of OOP is to facilitate *separation of concerns*, a phrase coined by Edsger Dijkstra in his 1974 essay titled, "On the Role of Scientific Thought."[1] In this essay he quotes himself:

> It is what I sometimes have called "the separation of concerns," which, even if not perfectly possible, is yet the only available technique for effective ordering of one's thoughts, that I know of. This is what I mean by "focussing one's attention upon some aspect": it does not mean ignoring the other aspects, it is just doing justice to the fact that from this aspect's point of view, the other is irrelevant. It is being one- and multiple-track minded simultaneously....

Object-oriented languages provide facilities to separate program concerns and focus on them independently, and, to encapsulate data abstractions and present them through well-defined interfaces. Complete object-oriented programs are constructed by separating the program's functionality into distinct classes, defining the interfaces for each class, and then establishing connections and interactions between components through their interfaces.

2.2   Classes and Objects

The primary unit of programming in object-oriented languages, such as SystemVerilog, is the *class*. A class contains data elements, called *members,* and tasks and functions, called *methods*. To execute an object-oriented program, you must *instantiate* one or more classes in a main routine and then call methods on the various objects. Although the terms class and object are sometimes used interchangeably, typically, the term *class* refers to a class declaration or an uninstantiated object, and the term *object* refers to an instance of a class.

To illustrate these concepts, below is an example of a simple class called `register`.

```
class register;
   local bit[31:0] contents;

   function void write(bit[31:0] d)
      contents = d;
```

---

1.   The complete text of Dijstra's essay is at http://www.cs.utexas.edu/users/EWD/ ewd04xx/EWD447.PDF

```
        endfunction

        function bit[31:0] read();
            return contents;
        endfunction
    endclass
```

This very simple class has one member, `contents`, and two methods, `read()` and `write()`. To use this class, you create objects by instantiating the class and then call the object's methods, as shown below

```
    module top;
        register r;
        bit[31:0] d;

        initial begin
            r = new();
            r.write(32'h00ff72a8);
            d = r.read();
        end
    endmodule
```

The `local` attribute on class member `contents` tells the compiler to strictly enforce the boundaries of the class. If you try to access `contents` directly, the compiler issues an error. You can only access contents through the publicly available read and write functions. This kind of access control is important to guarantee no dependencies on the internals of the class and thus enable the class to be reused.

You can use classes to create new data types, such as our simple `register`. Using classes to create new data types is an important part of OOP. You can also use them to encapsulate mathematical computations or to create dynamic data structures, such as stacks, lists, queues, and so forth. Encapsulating the organization of a data structure or the particulars of a computation in a class makes the data structure or computation highly reusable.

As a more complete example, let's look at a useful data type, the pushdown stack. A stack is a LIFO (last in first out) structure. Items are put into the stack with `push()`, and items are retrieved from the stack with `pop()`. `pop()` returns the last item pushed and removes it from the data structure. The internal member `stkptr` keeps track of the top of the stack. The item it points to is the top, and everything below it (that is, with a smaller index) is lower in the stack. Below is a basic implementation of a stack in SystemVerilog.

```
    43    class stack;
    44
```

```
45        typedef bit[31:0] data_t;
46        local data_t stk[20];
47        local int stkptr;
48
49        function new();
50          clear();
51        endfunction
52
53        function bit pop(output data_t data);
54
55          if(is_empty())
56            return 0;
57
58          data = stk[stkptr];
59          stkptr = stkptr - 1;
60          return 1;
61
62        endfunction
63
64        function bit push(data_t data);
65
66          if(is_full())
67            return 0;
68
69          stkptr = stkptr + 1;
70          stk[stkptr] = data;
71          return 1;
72
73        endfunction
74
75        function bit is_full();
76          return stkptr >= 19;
77        endfunction
78
79        function bit is_empty();
80          return stkptr < 0;
81        endfunction
82
83        function void clear();
84          stkptr = -1;
85        endfunction
86
87        function void dump();
88
89          $write("stack:");
90          if(is_empty()) begin
91            $display("<empty>");
92            return;
93          end
94
95          for(int i = 0; i <= stkptr; i = i + 1) begin
96            $write(" %0d", stk[i]);
97          end
```

```
98
99          if(is_full())
100             $write(" <full>");
101         $display("");
102
103     endfunction
104   endclass
file: 02_intro_to_OOP/01_stack/stack.sv
```

The class `stack` encapsulates everything there is to know about the stack data structure. It contains an *interface* and an *implementation* of the interface. The interface is the set of methods that you use to interact with the class. The implementation is the behind-the-scenes code that makes the class operate. The interface to our stack contains the following methods:

```
function new();
function bit pop(output DATA data);
function bit push(DATA data);
function bit is_full();
function bit is_empty();
function void clear();
function void dump();
```

There is no other way to interact with `stack` than through these methods. There are also two data members of the class, `stk` and `stkptr`, that represent the actual stack structure. However, these two members are local, which means that the compiler will disallow any attempts to access them from outside the class. By preventing access to the internals of the data structure from outside, we can make some guarantees about the state of the data. For example, `push()` and `pop()` can rely on the fact that `stkptr` is correct and points to the top of the stack. If it were possible to change the value of `stkptr` by means other than using the interface functions, then `push()` and `pop()` would have to resort to additional time-consuming and possibly unreliable checks to determine the validity of `stkptr`.

The implementation of the interface occurs inline. The class declaration contains not only the interface definition, but also the implementation of each of the interface functions. Both C++ and SystemVerilog allow the implementation to be separate from the interface. Separating the interface and the implementation is an important concept. Programmers writing in C++ can use header files to capture the interface and `.cc` (or `.cpp` or whatever the compiler uses) to hold the implementation.

There are some important by-products of enforcing access through class interfaces. One is reusability. We can more easily reuse classes whose interfaces are well defined and well explained than those whose interfaces are fuzzy. Another important by-product of enforcing access through class interfaces is reliability. The authors of the class can guarantee certain invariants (for example, stkptr is less than the size of the available stk array) when they know that users will not modify the data other than by the means provided. In addition, users can expect the state of the object to be predictable when they adhere to the interface. Clarity is another by-product. An interface can describe the entire semantics of the class. The object will do nothing other than execute the operations available through the interface. This makes it easier for those who use the class to understand exactly what it will do.

2.3   Object Relationships

The true power of OOP becomes apparent when objects are connected in various relationships. There are many kinds of relationships that are possible. We will consider two of the most fundamental relationships HAS-A and IS-A.

2.3.1   HAS-A

HAS-A refers to the concept of one object contained or owned by another. The HAS-A relationship is represented by members. In our stack class, for example, the stack HAS-A stack pointer (stkptr) and stack array. Those are primitive data types, not classes, but the same concept of HAS-A applies. In SystemVerilog you can create HAS-A relationships between classes with references or pointers. The figure below illustrates the underlying memory model for a HAS-A relationship. Object A contains a reference or a pointer to object B.

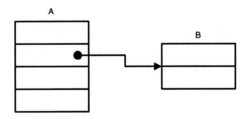

Figure 2-1 HAS-A Relationship

The Unified Modeling Language (UML) is a graphical language for representing systems, particularly the relationships between objects in those systems. The UML for a HAS-A relationship is expressed with a line between objects and a filled-diamond arrowhead, as in the diagram below.

Figure 2-2 UML for a HAS-A Relationship

Object A owns an instance of object B. Coding a HAS-A relationship in SystemVerilog involves instantiating one class inside another or in some other way providing a handle to one class that is stored inside another.

```
class B;
endclass

class A;
   local B b;
   function new();
      b = new();
   endfunction
endclass
```

class A contains a reference to class B. The constructor for class A, function new(), calls new() on class B to create an instance of it. The member b holds a reference to the newly created instance of B.

2.3.2 IS-A

The IS-A relationship is most often referred to as *inheritance*. A new class is *derived* from a previously existing object and *inherits* its characteristics. Objects created with inheritance are composed using IS-A. The derived object is considered a sub-class or a more specialized version of the parent object.

To illustrate the notion of inheritance, Figure 2-3 uses a portion of the taxonomy of mammals.

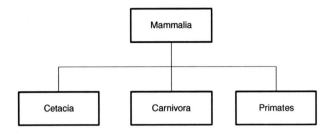

Figure 2-3 IS-A Example: Mammal Taxonomy

Animals that are members of the cetacia, carnivora, or primate orders are mammals. These very different kinds of creatures share the common traits of mammals. Yet cetacia (whales, dolphins), carnivora (dogs, bears, raccoons), and primates (monkeys, humans) each have their distinct and unmistakable characteristics. To use OO terminology, a bear IS-A carnivore and a carnivore IS-A mammal. In other words, a bear is composed of attributes of both mammals and carnivores plus additional attributes that distinguish it from other carnivores.

To express IS-A using UML, we draw a line between objects with an open arrow head pointing to the base class. Traditionally, we draw the base class above the derived classes, and the arrows point upward, forming an inheritance tree (or a directed acyclic graph that can be implemented in languages, such as C++, that support multiple inheritance).

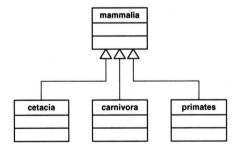

Figure 2-4 UML for IS-A Relationship

When composing two objects together in a computer program using inheritance, the new derived object contains characteristics of the parents and usually includes additional characteristics. The figure below illustrates the underlying memory model for an IS-A composition. In the example, the class B is *derived* from A.

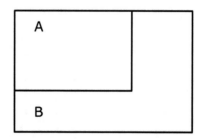

Figure 2-5 Example of IS-A Relationship

SystemVerilog uses the keyword extends to identify an inheritance relationship between classes:

```
class A;
   int i;
   float f;
endclass

class B extends A;
   string s;
endclass
```

Class B is derived from A, so it contains all the attributes of A. Any instance of B not only contains the string s, but also the floating point value f and the integer i.

2.4   Virtual Functions and Polymorphism

One of the reasons for composing objects through inheritance is to establish different behaviors for the same operation. In other words, the behavior defined in a derived class overrides behavior defined in a base class. The means to do this is through *virtual functions*. A virtual function is one that can be overridden in a derived class. Consider the following generic packet class.

```
class generic_packet;
   addr_t src_addr;
   addr_t dest_addr;
   bit m_header [];
   bit m_trailer []'
   bit m_body [];

   virtual function void set_header();
```

```
   virtual function void set_trailer();
   virtual function void set_body();
endclass
```

It has three virtual functions to set the contents of the packet. Different kinds of packets require different kinds of contents. We use generic_packet as a base class and derive different kinds of packets from it.

```
class packet_A extends generic packet;
   virtual function void set_header();
   endfunction
   virtual function void set_trailer();
   endfunction
   virtual function void set_body();
   endfunction
endclass

class packet_B extends generic_packet;
   virtual function void set_header();
   endfunction
   virtual function void set_trailer();
   endfunction
   virtual function void set_body();
   endfunction
endclass
```

Both packet_A and packet_B may have different headers and trailers and different payload formats. The knowledge about how the parts of the packet are formatted is kept locally inside the derived packet classes. The virtual functions set_header(), set_trailer(), and set_body() are implemented differently in each subclass based on the packet type. The base class generic_packet establishes the organization of the class and the types of operations that are possible, and the derived classes can modify the behavior of those operations.

Virtual functions are used to support *polymorphism*: multiple classes that can be used interchangeably, each with different behaviors. For example, some processing of packets may not need to know what kind of packet is being processed. The only information necessary is that the object is indeed a packet; that is, it is derived from the base class. Another way to say that is, the the current packet is related to the base class packet via the IS-A relationship. Virtual functions are the mechanism by which we can code alternate behaviors for different variations of a packet.

To look a little deeper at how virtual functions work, let's consider three classes related to each other by the IS-A relationship.

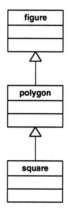

Figure 2-6 Three Classes Related with IS-A

figure is the base class; polygon is derived from figure; square is derived from polygon. Each class has two functions, draw(), which is virtual, and compute_area(), which is non-virtual. The following sample shows the SystemVerilog code:

```
38
39    class figure;
40
41       virtual function void draw();
42          $display("figure::draw");
43       endfunction
44
45       function void compute_area();
46          $display("figure::compute_area");
47       endfunction
48
49    endclass
50
51    class polygon extends figure;
52
53       virtual function void draw();
54          $display("polygon::draw");
55       endfunction
56
57       function void compute_area();
58          $display("polygon::compute_area");
59       endfunction
60
61    endclass
```

```
62
63    class square extends polygon;
64
65       virtual function void draw();
66          $display("square::draw");
67       endfunction
68
69       function void compute_area();
70          $display("square::compute_area");
71       endfunction
72
73    endclass
file: 02_intro_to_OOP/03_virtual/virtual.sv
```

Each function prints out its fully qualified name in the form class_name::function_name. We can write a simple program that calls each of these functions to understand how the virtual functions are bound.

```
75    program top;
76       figure f;
77       polygon p;
78       square s;
79
80       initial begin
81          s = new();
82          f = s;
83          p = s;
84
85          p.draw();
86          p.compute_area();
87          f.draw();
88          f.compute_area();
89          s.draw();
90          s.compute_area();
91       end
92    endprogram
file: 02_intro_to_OOP/03_virtual/virtual.sv
```

The following shows what happens when we run this program:

```
square::draw
polygon::compute_area
square::draw
figure::compute_area
square::draw
square::compute_area
```

First we create s, a square, and then we assign it to f and p. The immediate base class of square is polygon and the base class of polygon is figure. From the printed output, we can conclude that the functions are bound according to the following table:

p.draw()	square::draw()
p.compute_area()	polygon::compute_area()
f.draw()	square::draw()
f.compute_area()	figure::compute_area()
s.draw()	square::draw()
s.compute_area()	square::compute_area()

In all cases, compute_area() was bound to the particular compute_area() function specified by the type of the reference that called it—p is a reference to a polygon, thus polygon::compute_area() is bound. This is because compute_area() is non-virtual. The compiler can easily determine which version of the function to call simply based on the type of the object.

Because draw() is virtual, it is not always possible for the compiler to determine which function to call. The decision is made at run time using a *virtual table*, a table of function bindings. A virtual table is used to bind functions whose bindings cannot be entirely determined at compile time. A good reference for learning more about how virtual tables work is *Inside the C++ Object Model* by Stanley B. Lippman.

Notice that even though p is a polygon, the call to p.draw() results in square::draw() being called not polygon::draw(), as you might expect. The same thing happens with f—f.draw() is bound to square::draw(). The object we originally instantiated is a square, and even though we assign handles of different types, the fact that it is a square is not forgotten. This works only because square is derived from polygon, which in turn is derived from figure, and because draw() is declared as virtual. A compile time error about type incompatibility occurs if you try to assign s to p and s is not derived from p.

2.5   Generic Programming

Recall that object-oriented languages provide facilities to separate program concerns and focus on them independently. An implication of separating concerns is that each concern is represented only once. Duplicating code violates the principle. In practice, many problems are quite similar, and their solution requires code that is similar, but not identical. Intuitively, we want to take advantage of code similarity to write code that can be used in as many situations as possible. This intuition leads us to writing generic code, code that is highly parameterized so that it can be easily reused in a wide variety of situations.

Details of generic code are supplied at compile time or run time instead of hard coding them. Any code that has parameters, such as function calls, can be considered generic, but the term is usually reserved for code built around templates (in C++) or parameterized classes (in SystemVerilog). Making programs generic is consistent with the OOP goal of separating concerns. Thus OOP languages provide facilities for building generic code.

A parameterized class is one that (obviously) has parameters. The syntax in SystemVerilog for identifying parameters is a pound sign (#) in the class header followed by a parenthesized list of parameters. As an example, consider the following parameterized class:

```
class param #(type T=int, int R=16);
endclass
```

This class has two parameters, T, which is a type parameter and R, which is an integer parameter. Instances of a parameterized class with specific values for the parameters create specializations, that is, versions of the code with the parameters applied.

```
param #(real, 29) z;
param #(int unsigned, 12) q;
```

The above declarations create specializations of the parameterized class param. The class name and parameters identify specializations. Thus, specializations are in fact, unique types. The compiler will not allow you to assign q to z, or vice versa, because they are objects of different types.

type parameters allow you to write type-independent code, code whose data structures and algorithms can operate on a wide range of data types. For example:

```
class maximizer #(type T=int);
   function T max(T a, T b);
      if( a > b )
           return a;
      else
           return b;
   endfunction
endclass
```

The parameterized class maximizer has a function max() that returns the maximum of two values. The max algorithm is the same no matter the type of the comparison objects. In this case, the only restriction is that the objects be comparable with the greater than (>) operator.

Classes cannot be meaningfully compared using the greater-than operator, so a different version of maximizer is necessary to deal with classes. To make a version of maximizer that will return the largest of two class objects, we must define a method in each class that will compare objects.

```
class maximizer #(type T=int);
   function T max( T a, T b);
      if( a.comp(b) > 0 )
            return a;
      else
            return b;
   endfunction
endclass
```

This presumes that the type parameter T is really a class, not a built-in type, such as int or real. Further, it presumes that T has a function called comp(), which is used to compare itself with another instance. The OVM library contains a parameterized component called ovm_in_order_comparator#(T), which is used to compare streams of transactions. It has two variants, one for comparing streams of built-in types, and one for comparing streams of classes. The reason we need two in-order comparator classes is exactly the same reason we need two maximizers—SystemVerilog does not support operators that can operate on either classes or built-in types.

2.5.1   Generic Stack

Our stack is not particularly generic. It has a fixed stack size of 20, and the data type of the items kept on the stack is fixed to be int. Below is a more generic form of stack that changes these fixed characteristics to parametrized characteristics.

```
53      class stack #(type T = int);
54
55         local T stk[];
56         local int stkptr;
57         local int size;
58         local int tp;
59
60         function new(int s = 20);
61            size = s;
62            stk = new [size];
63            clear();
64         endfunction
65
```

```
66       function bit pop(output T data);
67
68         if(is_empty())
69           return 0;
70
71         data = stk[stkptr];
72         stkptr = stkptr - 1;
73         return 1;
74
75       endfunction
76
77       function bit push(T data);
78
79         if(is_full())
80           return 0;
81
82         stkptr = stkptr + 1;
83         stk[stkptr] = data;
84         return 1;
85
86       endfunction
87
88       function bit is_full();
89         return stkptr >= (size - 1);
90       endfunction
91
92       function bit is_empty();
93         return stkptr < 0;
94       endfunction
95
96       function void clear();
97         stkptr = -1;
98         tp = stkptr;
99       endfunction
100
101      function void traverse_init();
102        tp = stkptr;
103      endfunction
104
105      function int traverse_next(output T t);
106        if(tp < 0)
107          return 0; // failure
108
109        t = stk[tp];
110        tp = tp - 1;
111        return 1;
112
113      endfunction
114
115      virtual function void print(input T t);
116        $display("print is unimplemented");
117      endfunction
118
```

```
119      function void dump();
120
121        T t;
122
123        $write("stack:");
124        if(is_empty()) begin
125          $display("<empty>");
126          return;
127        end
128
129        traverse_init();
130
131        while(traverse_next(t)) begin
132          print(t);
133        end
134        $display();
135
136      endfunction
137
138    endclass
file: 02_intro_to_OOP/02_generic_stack/stack.sv
```

The generic stack class is parameterized with the type of the stack object. The parameter T contains a type. In this case, T can be either a class or a built-in type because we are not using operators directly on objects of type T. Any place in the class where we previously used int as the stack type, we now use T. For example, push() now takes an argument of type T. Class parameters, such as T, are compile-time parameters, meaning the value is established at compile time. To specialize stack#(T), we instantiate it with a specific value for the type. For example:

```
stack #(real) real_stack;
```

This statement creates a specialization of stack that uses real as the type of object on the stack.

The size of the stack is no longer fixed at 20. We use a dynamic array to store the stack, whose size is specified as a parameter to the constructor. Unlike T, the argument size is a run-time parameter—its value is specified when the program runs. This lets us create multiple stacks, each with a different size.

```
stack #(real) big_stack;
stack #(real) little_stack;

...

big_stack = new(2048);
little_stack = new(6);
```

big_stack and little_stack are of the same type. They use the same specialization of stack#(T). However, they are each instantiated with different size parameters.

In making stack generic, we made another change. We replaced dump() with traverse_init() and traverse_next(). dump() relies on the type of the stack elements, which is not known until compile time. We need to be able to traverse the stack and format each element no matter what the element type is. It could be an int, or it could be a complex class with multiple members. We don't know what it will be. To keep stack#(T) generic, we must resist all temptation to establish any reliance on the type of the stack elements.

Whereas dump() will run through the stack elements and print them in order, traverse_init() sets an internal traversal pointer (tp) to point to the top of the stack, and traverse_next() hands the current element (as pointed to by tp) back to the caller and decrements tp. The stack maintains some state information about the traversal. The state information is reset when traverse_init() is called.

By making stack#(T) generic, removing reliance on hardcoded types and sizes, we have made this component highly reusable.

2.6   Classes and Modules

Interestingly, HDLs, such as Verilog and VHDL, though not considered object-oriented languages, are built around concepts quite similar to classes and objects. Module instances in Verilog, for example, are objects, each with its own data space and set of tasks and functions. Just like objects in OO programs, each instance of a module is an independent copy. All instances share the same set of tasks and functions and the same interfaces, but the data contained inside each one is independent from all other instances. Modules are controlled by their interfaces. Verilog modules do not support inheritance (that is, the ability to form IS-A relationships) or type parameterization, and they are static, which makes them unsuitable for true OOP.

The similarity between classes and modules opens up an opportunity for us to use class objects in a hardware context. We can create verification components as instances of classes, giving us the flexibility of classes along with the connection to hardware elements. The designers of SystemVerilog have capitalized on this relationship when extending Verilog with classes, providing the capability for a class to work a lot like modules.

The table below compares features of classes in Verilog, SystemVerilog, and C++.

Feature	Verilog Modules	C++ Classes	SystemVerilog Classes
local data space	yes	yes	yes
function interface	kind of	yes	yes
port interface	yes	no	yes
inheritance	no	yes/multiple	yes/single
type parameterization	no	yes	yes
dynamic	no	yes	yes

The SystemVerilog feature that makes this possible is the *virtual interface*. A virtual interface is a reference to an interface (here we refer to the SystemVerilog interface construct). We can write a class containing references to items inside an interface that doesn't yet exist (that is, it isn't instantiated). When the class is instantiated, the virtual interface is connected to a real interface. This makes it possible for a class object to both drive and respond to pin activity. SystemC modules are implemented as classes and allow for pins to be in the port list, providing the same sort of structure.

HDLs, such as Verilog and VHDL, lack many OOP facilities, and thus are not well suited for building testbenches. The fundamental unit of programming in most HDLs is the module, which is a static object. Modules come into existence at the very beginning of the program and persist unmodified until the program completes. They are syntactically static as well—the syntactic means to modify a module to create a variant are limited. Verilog allows you to parameterize scalar values, but not types. Often you are reduced to cutting and pasting code, then making local modifications. If you have ten different variations you need in a particular design, you must paste ten copies in appropriate locations and then locally modify each one. Should the template module change (the one that you pasted around to create the variants), you'll have to locate each instance and make those same changes in each one. This process is not all that different from what our assembly language programmers had to do fifty years ago.

Sidebar: Simula 67

The relationship between class objects and hardware simulation has been around for quite some time. Simula 67,[1] one of the earliest OOP languages, was developed explicitly for the purpose of building discrete event models. Simula 67 has the notion of class objects and a simulation kernel. It even has a kind of PLI for connecting in external Fortran programs. Simula provides DETACH and RESUME keywords, which allow processes to be spawned and reconnected, sort of a fork/join. It has a special built-in class called SIMULATION, which provides event list features.

Even though the terms object and object-oriented are not used at all in Simula 67, all modern object-oriented programs can trace their lineage to this early programming language. Discrete event simulation languages also can trace their genesis to Simula 67. For many, bringing together the ideas of OOP and hardware simulation seems new; but in fact, the two ideas were born together and only later parted ways. Using OOP with a discrete event simulator brings us full circle.

According to Ole-Johan Dahl and Kristen Nygaard, Department of Informatics, University of Oslo:[2]

> Simula 67 still is being used many places around the world, but its main impact has been through introducing one of the main categories of programming, more generally labelled object-oriented programming. Simula concepts have been important in the discussion of abstract data types and of models for concurrent program execution, starting in the early 1970s. Simula 67 and modifications of Simula were used in the design of VLSI circuitry (Intel, Caltech, Stanford). Alan Kay's group at Xerox PARC used Simula as a platform for their development of Smalltalk (first language versions in the 1970s), extending object-oriented programming importantly by the integration of graphical user interfaces and interactive program execution. Bjarne Stroustrup started his development of C++ (in the 1980s) by bringing the key concepts of Simula into the C programming language. Simula has also inspired much work in the area of program component reuse and the construction of program libraries.

1. Lamprecht, Gunther, "Introduction To Simula 67," Vieweg, 1983
2. http://heim.ifi.uio.no/~kristen/FORSKNINGSDOK_MAPPE/F_OO_start.html

2.7   OOP and Verification

Building an object-oriented program and building a testbench are not very different things. A testbench is a network of interacting components. OOP deals with defining and analyzing networks of interacting objects. Objects can be related through IS-A or HAS-A, and they communicate through interfaces. OOP just naturally fits the problem of building testbenches.

Languages such as SystemC/C++ and SystemVerilog, which do provide OOP facilities, are better suited for testbench construction than HDLs, such as Verilog and VHDL. Using dynamic classes, parameterized classes, inheritance, and parameterized constructors, you can build components that are flexible, reusable, and robust. Spending a little extra time to build a generic component can result in a large productivity gain when that component is reused in different ways in many places.

3

Transaction-Level Modeling

The process of designing an electronic system involves taking abstract ideas and successively replacing the abstractions with concrete details until you reach a representation that can be manufactured in silicon. Since the advent of the digital integrated circuit, the electronic design community has carefully defined and codified abstractions, beginning with switches and gates, to provide media in which designs are rendered. RTL is an example of an abstraction medium commonly used to create designs. There are many tools based on the RTL abstraction that make it a convenient way to initiate the design and verification process.

However, as designs get larger and more complex, it becomes increasingly convenient to represent them using abstractions higher than RTL. The transaction level is becoming popular for creating the first incarnation of a design that can be simulated and analyzed.

This chapter introduces the fundamental concepts of transaction-level modeling (TLM). Transaction-level models consist of multiple processes communicating with each other by sending transactions back and forth through channels. This chapter illustrates these concepts with some producer-consumer pairs communicating through transaction-level interfaces.

3.1   Abstraction

In their book *System Design with SystemC*, Grötker *et al.*, discuss models of computation. They define a model of computation as having three components:

- A model of time

M. Glasser, *Open Verification Methodology Cookbook*, DOI: 10.1007/978-1-4419-0968-8_3,

- Methods of communication between concurrent processes
- Rules for process activation

RTL modeling uses a discrete model of time. Communication between processes is done using nets, and process activation occurs when an input net of a process changes its value.

In comparison, transaction-level models can be timed *or* untimed and use *channels* to communicate between processes. Instead of sending individual bits back and forth, the processes communicate by sending transactions to each other through function calls.[1] The world of TLM encompasses a range of models of computation with different time, communication, and process activation models. In each case, however, the contents of the communication are at a higher level of abstraction than individual bits. Thus, a transaction-level model is at a higher level of abstraction (it is more abstract) than an RTL model. Combining the notions of abstraction and models of computation, we can see that making an abstract model means abstracting time, data, and function. The following sections discuss these elements in detail.

Abstract time. The time abstraction in a simulator refers to how often the entire design state is consistent. Models that run in event-driven simulators (for example, logic simulators) use a discrete notion of time, meaning events happen at specific time points. Events usually (although not always) cause a process of some sort to be invoked. As more events occur in a simulation, more processes are invoked, and with more processes comes slower overall simulation runs. Abstracting time reduces the number of points where the design must be consistent and, therefore, the total number of events and process activations that must occur. For example, in an RTL model, every net must be consistent after every change. In cycle-accurate abstraction, the design must be consistent only on the clock edges, eliminating all the events that occur between clock edges. In a transaction-level model, the design state must be consistent at the end of each transaction, each of which might span many clock cycles.

Abstract data.   Data   refers   to   the   objects   communicated   between components. In RTL models, the data refers to individual bits that are passed via nets between components. In transaction-level models, data is in the form of transactions, heterogeneous structures that contain arbitrary collections of elements.

---

1. You could easily make the case that a transfer of a single bit is a transaction in the most general sense. And even though a bit might be considered a transaction, this discussion on transactions restricts the concept to cases involving higher levels of abstractions than bits.

Consider a packet in a communications device. At the lowest level of detail, the packet contains start and stop bits, a header, error correction information, payload size, payload, and a trailer. In a more abstract model, only the payload and size might be necessary. The other pieces of data are not necessary for the calculations being performed.

Abstract function. The function of a model is the set of all things it must do at each event. Abstracting function reduces that set or replaces it with simpler calculations. For example, in an ALU, you might choose to use the native multiplication operation supplied in your modeling language instead of coding the complete algorithm for a shift-and-add multiplier. The latter may be part of the implementation, but at the higher level, the details of the shift-and-add algorithm are unimportant. The primitives that are part of the language define how you can abstract function. In a gate-level language, for example, you build complex behaviors from gates. In an RTL language, you build behaviors around arithmetic and logical operations on registers. In TLM, you implement design functionality with function calls of arbitrary complexity.

For the purposes of functional verification, RTL is the lowest-level abstraction that we need to consider. Since synthesizers can effectively convert RTL to gates, we don't need to concern ourselves with lower levels of detail. Besides, anything lower gets into electrical issues that are beyond the scope of logic design.

3.2   Definition of a Transaction

To effectively talk about TLM in greater detail, we must step back and define transactions.

> *A transaction is a quantum of activity that occurs in a design bounded by time.*

This is the most general definition of a transaction. It says that a transaction is everything that occurs in a design (or a module or subsystem within a design) between two time points. While that is accurate, it is so general that it doesn't lead to practical application. A more useful definition is the following:

> *A transaction is a single transfer of control or data between two entities.*

This is the hardware-oriented notion of a transaction. When looking at a piece of hardware, you can easily identify entities between which control or data is transferred. In a bus-based design, reads and writes on a bus can be considered transactions. In a packet-based communication system, sending a packet is a transaction.

The following is a third definition:

> *A transaction is a function call.*

This definition is the software-oriented notion of a transaction. In a transaction-level model, activity is initiated by making function calls. The function call contains parameters that are "sent" to the called function, and the return value of the function contains data that is returned by the called function. The called function could block and cause time to pass (in a timed system) or it could return immediately.

3.3   Interfaces

Before we go into the details about how to build transaction-level models, we will first take a small detour to discuss interfaces. The term *interface* is used in several ways in OVM, each with a slightly different meaning. It's an unfortunate fact of history that the same word has come to mean so many different things. Most of the time you can understand the meaning from the context in which the term is used. The different uses in this book are the following:

- SystemVerilog interface
- Object interface
- DUT interface

SystemVerilog Interface. SystemVerilog provides a construct called an interface, which is one of the primary container objects from which you construct a design in SystemVerilog. We use virtual interfaces, which are essentially pointers to interfaces, to connect module-based hardware to class-based testbenches. The next chapter looks at the details of making that connection.

Object Interface. The publicly visible tasks and functions available on an object form its interface. There are two slight variations of this meaning of interface. One is straightfoward. Look at a class and determine what tasks and functions are available to the user of the class to operate it. That's its interface. The other variation is to refer to a base class that defines the set of tasks and functions available to operate the derived class. This meaning of interface is more typically used with object-oriented languages that support multiple inheritance, such as C++ or Java[1]. In those languages, you can establish a requirement that the derived class supply certain functionality by inheriting from an interface base class.

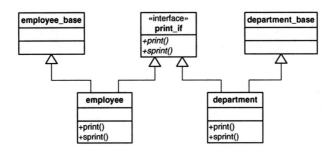

Figure 3-1 Interface Inheritance (with Multiple Inheritance)

The `print_if` interface specifies the prototypes for the print functions. Any class that inherits from `print_if` is then obliged to implement `print()` and `sprint()`. SystemVerilog does not support multiple inheritance, but it does support *pure virtual interfaces*. A pure virtual interface is an interface in this second context (a base class that defines a set of task and function prototypes) that has no implementations. A pure virtual version of our `print_if` would look as follows in SystemVerilog:

```
virtual class print_if;
   pure virtual function void print();
   pure virtual function string sprint();
endclass
```

Even though SystemVerilog does not support multiple inheritance, and OVM is built on SystemVerilog, it is important to understand pure virtual interfaces and interface inheritance because they are used heavily in OVM. In particular, TLM ports and exports are derived from an interface class called

---

1. Java doesn't support full multiple inheritance in the same manner as C++. It does support interface inheritance. This establishes the requirement that a class derived from an interface provide the specified functionality.

`tlm_if_base`. Later in this book, we will have more discussion on port construction.

DUT Interface. A piece of hardware is typically accessed through its interfaces. In this context, an interface is composed of the pins and protocol used to communicate to the device. For example, a device may have a USB interface.

3.4   TLM Idioms

This section reviews the basic means of transmitting a transaction between components. We'll examine put, get, and transport forms of transaction communication. These examples do not use the OVM library, as they are intended to illustrate the essential mechanics of transaction-level communication with minimal overhead. In the next section we'll look at a more complete example that uses the OVM library for communication.

3.4.1   Put

In a put configuration, one component sends transactions to another component. The operation is called a *put*. The *initiator* is the component that initiates the transfer, and the *target* is the component that receives the result. Using TLM nomenclature, we say that the initiator *puts* transactions to the target.

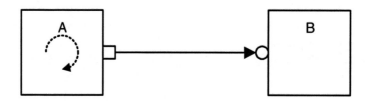

Figure 3-2 Put

Figure 3-2 indicates that A puts transactions to B. The initiator has a *port* drawn as a square box, and the target has an *export* drawn as a circle. The flow of control is from box to circle; that is, A will call B, which contains an implementation of the port methods. The arrow shows the direction of the data flow, and in this case, it indicates that data will move from A to B.

We can illustrate the code for these components with a `producer` and a `consumer`. The `producer` is the initiator and the `consumer` is the target. We must build these components in such a way that they do not know about each other *a priori*. To do that, we use a pure virtual interface to define the function that will be used to transmit data between the initiator and the target. First, let's take a look at the SystemVerilog version of the producer.

```
46    class producer;
47
48       put_if put_port;
49
50       task run();
51
52          int randval;
53
54          for(int i=0; i<10; i++)
55             begin
56                randval = $random %100;
57                $display("producer: sending    %4d", randval);
58                put_port.put(randval);
59             end
60
61       endtask
62
63    endclass : producer
file: 03_tlm/01_put/put.sv
```

The producer is a class, implying that it is created dynamically. It has two key elements, a run() task and a put_port. The run() task is a simple task that loops 10 times and puts 10 transactions. To keep things simple, our transactions are integers. In practice, a transaction can be an arbitrarily complex object such as a struct or a class.

To put transactions, the producer calls put() on the put_port. What is a put_port? It is not a port in the traditional Verilog sense. It is a reference to a put_if. What is a put_if? A put_if is the virtual interface class shared between the initiator (producer) and target (consumer).

```
39    virtual class put_if;
40       pure virtual task put(int val);
41    endclass : put_if
```

put_if is a class with a pure virtual task; meaning, the task has no implementation. Without an implementation of all of its tasks and functions, a virtual class cannot be instantiated by itself. It must be the base class of another class that is instantiated. In our case, the class derived from the pure virtual put_if is consumer.

```
68    class consumer extends put_if;
```

```
69        task put(int val);
70          $display("consumer: receiving %4d", val);
71        endtask : put
72     endclass : consumer
file: 03_tlm/01_put/put.sv
```

consumer contains an implementation of put(); the pure virtual task defined in put_if. The put() task implementation accepts the argument passed to it and prints it. put_if plays a pivotal role in connecting the producer to the consumer. A reference to it on the producer side, which we call a port, establishes the requirement that there must be an implementation of the functions and tasks in the interface to which this object will be bound. The consumer is derived from the interface and, therefore, must implement the pure virtual task satisfying the requirement.

The top-level module binds the producer to the consumer.

```
77     module top;
78
79        producer p;
80        consumer c;
81
82        initial begin
83           // instantiate producer and consumer
84           p = new();
85           c = new();
86           // connect producer and consumer
87           // through the put_if interface class
88           p.put_port = c;
89           p.run();
90        end
91     endmodule : top
file: 03_tlm/01_put/put.sv
```

Notice the assignment statement:

```
88           p.put_port = c;
```

It forms the linkage between the producer and the consumer. When new() is called on p to create a new instance of producer, the member put_port has no value. A run-time failure will occur if put_port.put() is called prior to the linkage assignment. Assigning c to p.put_port gives the port a reference to the consumer, which contains an implementation of the interface task put().

3.4.2 Get

The complement to put is *get*. In this arrangement, the initiator receives a transaction from the target. The flow of control is the same—from initiator to target—but the direction of the data flow is the opposite. The initiator gets a transaction from the target. In this case, the consumer is the initiator and the producer is the target. The consumer initiates a call to the producer to retrieve a transaction.

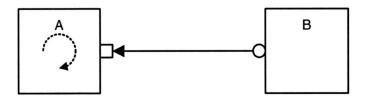

Figure 3-3 Get Configuration

Figure 3-3 is very similar to Figure 3-2. The only difference is that here the arrow points from the target to the initiator instead of the other way around. This indicates that the data flows from the target to the initiator. The following is the SystemVerilog consumer (initiator).

```
62    class consumer;
63
64      get_if get_port;
65
66      task run();
67        int randval;
68        for(int i=0; i<10; i++)
69        begin
70          get_port.get(randval);
71          $display("consumer: receiving %4d", randval);
72        end
73      endtask
74    endclass
file: 03_tlm/02_get/get.sv
```

The consumer has a task, run(), which iterates 10 times to get 10 transactions. Like the producer in the put example, the consumer here has a port. Also like the put example, the port is a reference to a pure virtual interface, in this case it is called get_if.

```
41    virtual class get_if;
42      pure virtual task get(output int t);
```

```
43      endclass : get_if
file: 03_tlm/02_get/get.sv
```

get_if is a pure virtual interface class that defines the task get(). The target
(producer) is constructed in a similar fashion to the target in the put example.
It contains an implementation of the interface task. This producer produces a
random value between 0 and 99.

```
48      class producer extends get_if;
49
50        task get(output int t);
51          int randval;
52          randval = $random % 100;
53          $display("producer: sending    %4d", randval);
54          t = randval;
55        endtask
56
57      endclass : producer
file: 03_tlm/02_get/get.sv
```

The connection at the top level will look very familiar.

```
79      module top;
80
81        producer p;
82        consumer c;
83
84        initial begin
85          // instantiate producer and consumer
86          p = new();
87          c = new();
88          // connect producer and consumer through the get_if
89          // interface class
90          c.get_port = p;
91          c.run();
92        end
93      endmodule : top
file: 03_tlm/02_get/get.sv
```

After creating instances of the producer and consumer by calling new(), the
two components are connected using a linkage assignment.

3.4.3 Transport

Transport is a bidirectional interface. The interface provides for transactions
to be sent from the initiator to the target and from the target back to the

initiator. Typically, we use this arrangement to model request-response protocols. When talking about components with bidirectional interfaces, we use the terms *master* and *slave* instead of initiator and target.

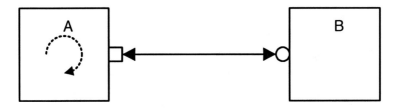

Figure 3-4 Bidirectional Transport Configuration

The master (A) does both a put and a get in a single function call. As we saw in previous sections, put () and get () tasks each take one argument, the argument they are putting or getting. However, the transport () task takes two arguments, a request and a response. It sends the request and returns with a response. The slave (B) accepts the request and replies with a response.

Let's first look at the pure virtual interface.

```
37    virtual class transport_if;
38       pure virtual task transport(input int request,
39                                    output int response);
40    endclass : transport_if
file: 03_tlm/03_transport/transport.sv
```

The interface contains a single task, transport (), which takes two arguments: a request that is passed to the target and a response that is returned back to the initiator.

The master calls transport (), creates a request, and sends it to the slave via transport. It processes the response that is returned.

```
45    class master;
46
47       transport_if port;
48
49       task run();
50
51          int request;
52          int response;
53
54          for(int i=0; i<10; i++)
55             begin
```

```
56              request = $random % 100;
57              $display("master: sending   request   %4d",
58                      request);
59              port.transport(request, response);
60              $display("master: receiving response %4d",
61                      response);
62          end
63
64      endtask
65  endclass : master
file: 03_tlm/03_transport/transport.sv
```

The slave implements the transport() task. In our example, it does some trivial processing of the request to create a response.

```
70  class slave extends transport_if;
71
72      task transport(input int request, output int response);
73              $display("slave:  receiving request  %4d",
74                      request);
75              response = -request;
76              $display("slave:  sending   response %4d",
77                      response);
78          endtask
79
80  endclass
file: 03_tlm/03_transport/transport.sv
```

The top-level linkage between master and slave works the same way the put and get examples work.

```
85  module top;
86
87    master m;
88    slave s;
89
90    initial begin
91      // instantiate the master and slave
92      m = new();
93      s = new();
94
95      // connect the master and slave through
96      // the port interface
97      m.port = s;
98      m.run();
99    end
100
101  endmodule : top
file: 03_tlm/03_transport/transport.sv
```

The linkage assignment makes the connection between the master and the slave. After the assignment completes, the master can use the connection to directly call functions in the slave.

3.4.4   Blocking vs. Nonblocking

The interfaces we have looked at so far are *blocking*. That means that the functions and tasks block execution until they complete. They are not allowed to fail. There is no mechanism for any blocking call to terminate abnormally or otherwise alter the flow of control. They simply wait until the request is satisfied. In a timed system, this means that time may pass between the time the call was initiated and the time it returns.

In the put configuration, we have two components, producer and consumer. The producer generates a random number and sends it to the consumer via put(). Before put() is called, there is no activity in the consumer. The call to put() causes activity in the consumer, which prints the value of the argument. During the time that the consumer is active, the producer is waiting. This is the nature of a blocking call. The caller must wait until the call finishes to resume execution.

Now contrast that description with a *nonblocking* call. A nonblocking call returns *immediately*. The semantics of a nonblocking call guarantee that the call returns in the same delta cycle in which it was issued, that is, without consuming any time, not even a single delta cycle.

The pure virtual interface that connects the nonblocking slave to the master looks much like the other pure virtual interfaces we've seen. The significant difference is that the nb_get() function returns a status value instead of a transaction.

```
41    virtual class get_if;
42      pure virtual function int nb_get(output int t);
43    endclass : get_if
file: 03_tlm/04_nonblocking/nbget.sv
```

The master (consumer) must check the status return from nb_get() to determine whether the function successfully completed. Notice also that we've introduced time into the model. The consumer checks every 4 ns to see if a value is available.

```
78    class consumer;
```

```
79
80        get_if get_port;
81
82        task run();
83           int randval;
84           int ok;
85
86           for(int i=0; i<20; i++)
87             begin
88               #4;
89               if(get_port.nb_get(randval))
90                 $display("%t: consumer: receiving %4d", $time,
randval);
91               else
92                 $display("%t: consumer: no randval", $time);
93             end
94        endtask
95    endclass
file: 03_tlm/04_nonblocking/nbget.sv
```

The producer is organized as a function and a task. The task will be forked
(spawned) to run as a continuous process. It generates new random values
that the consumer will grab. However, each random value is only available
for 2 ns out of a 7 ns cycle. The function is an implementation of nb_get that
returns the value generated periodically by the run() task.

```
48    class producer extends get_if;
49
50        int randval = 0;
51        int rand_avail = 0;
52
53        function int nb_get(output int t);
54           if(rand_avail)  begin
55             $display("%t: producer: sending    %4d",
56                        $time, randval);
57             t = randval;
58             return 1;
59           end
60           return 0;
61        endfunction
62
63        task run();
64           forever begin;
65             #5;
66             randval = $random % 100;
67             rand_avail = 1;
68             #2;
69             rand_avail = 0;
70           end
71        endtask
```

```
72
73     endclass : producer
file: 03_tlm/04_nonblocking/nbget.sv
```

When we run the example, we see that not every nb_get() call succeeds.

```
 4: consumer: no randval
 8: consumer: no randval
12: producer: sending    -99
12: consumer: receiving  -99
16: consumer: no randval
20: producer: sending    -39
20: consumer: receiving  -39
24: consumer: no randval
28: producer: sending     -9
28: consumer: receiving   -9
32: consumer: no randval
36: consumer: no randval
40: producer: sending     57
40: consumer: receiving   57
44: consumer: no randval
48: producer: sending    -71
48: consumer: receiving  -71
52: consumer: no randval
56: producer: sending    -14
56: consumer: receiving  -14
60: consumer: no randval
64: consumer: no randval
68: producer: sending     29
68: consumer: receiving   29
72: consumer: no randval
76: producer: sending     18
76: consumer: receiving   18
80: consumer: no randval
```

The blocking get configuration had only one process—the consumer that continually made requests to the producer to send a new value. The nonblocking variant has two processes: the consumer regularly polls the producer to see if it has a value to grab, and the producer generates new values asynchronously with respect to the consumer. Our nonblocking producer makes a random value available every 7 ns. It waits 5 ns and then generates a new value, and the new value is valid for 2 ns. The flag rand_avail is set when a valid random value is available and cleared when none is available.

The implementation of nb_get() for this example must check rand_avail to see if there is indeed something to send. If not, it returns a 0 to indicate that the request failed. If there is something available, then it sends it and returns a 1 to indicate success.

Blocking interfaces are useful for operating two components synchronously. Blocking calls wait until the requested operation completes, no matter how long that might take. On the other hand, nonblocking interfaces are useful for communicating asynchronously. They do not wait and can be used to poll targets, as in the example shown.

3.5   Isolating Components with Channels

The previous section discussed simple mechanisms for moving a transaction between two processes. In each, the initiator and target were tightly synchronized by the transaction interface task call. In this section, we examine the case where the initiator and target are more loosely coupled. The decoupling is possible using a channel, in this case a FIFO, to manage the synchronization between the initiator and the target, rather than relying on the two components to synchronize themselves. Here we have two components, an initiator A and a target B, plus a FIFO connecting the two components.

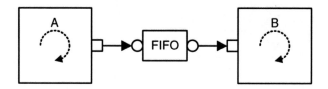

Figure 3-5   Two Components Isolated with a FIFO

In the previous examples, one component had a port and the other an export. The component with the port makes calls to the component with the export. Here both A and B have ports. Instead of the initiator calling the target directly, now we have both the initiator and the target calling the FIFO channel. The channel provides the functions required by both the initiator and the target.

The initiator uses a blocking put() to send transactions to the FIFO, and the target uses a blocking get() to retrieve transactions from the FIFO. The FIFO buffers the transactions and serves as a synchronizer. The initiator can continue putting transactions into the FIFO until it is full. Since the initiator uses a blocking put(), the initiator process will block when the FIFO is full. Likewise, the target uses a blocking get() and will block when the FIFO is empty. Essentially, the producer in this example is like the producer in the blocking put example, and this consumer is like the consumer in the blocking get example. The FIFO replaces the target and provides the tasks necessary to satisfy the interface requirements created by the ports on the producer and consumer.

Let's look at the code. This is the first example that uses the OVM library. The OVM library includes a FIFO, called tlm_fifo, which is a parameterized class with a variety of interfaces to support blocking and nonblocking operations.

This producer looks a lot like the producer in the blocking put example. It has a process, run(), that loops 10 times, generating 10 random values and sending them to the target via the put_port.

```
42    class producer extends ovm_component;
43
44      ovm_blocking_put_port#(int) put_port;
45
46      function new(string name, ovm_component p = null);
47        super.new(name,p);
48        put_port = new("put_port", this);
49      endfunction
50
51      task run();
52
53        int randval;
54        string s;
55
56        for(int i = 0; i < 10; i++)
57          begin
58            randval = $random % 100;
59            $sformat(s, "sending    %4d", randval);
60            ovm_report_info("producer", s);
61            put_port.put(randval);
62          end
63        global_stop_request(); // OK, we're done now
64      endtask
65
66    endclass
file: 03_tlm/05_fifo/fifo.sv
```

There are two new things to notice. First, the component is derived from ovm_component, which is a base class in the OVM library that provides essential services for components. It allows components to be connected into the hierarchy of named components, and it provides process control for the run task. The run task is forked at startup and can be suspended or resumed at will.

The other thing to notice is how put_port is declared. In our simple examples above, we created our own pure virtual interface to connect the initiator to the target. The OVM library supplies a collection of port and export objects, which are wrappers around pure virtual interface references. The port and export objects, which are themselves named components, provide a connect() function for establishing associations between ports and exports. This is a nicer use model compared to using assignment statements.

The consumer is not much different than the consumer in the blocking get example.

```
71    class consumer extends ovm_component;
72
73      ovm_blocking_get_port#(int) get_port;
74
75      function new(string name, ovm_component p = null);
76        super.new(name,p);
77        get_port = new("get_port", this);
78      endfunction
79
80      task run();
81
82        int val;
83        string s;
84
85        forever
86          begin
87            get_port.get(val);
88            $sformat(s, "receiving %4d", val);
89            ovm_report_info("consumer", s);
90          end
91
92      endtask
93
94    endclass
file: 03_tlm/05_fifo/fifo.sv
```

To connect the producer, consumer, and fifo, we use an environment. An environment serves as the top of the hierarchy of named components, and it orchestrates the hierarchy construction and testbench execution.

```
99    class env extends ovm_component;
100     producer p;
101     consumer c;
102     tlm_fifo #(int) f;
103
104     function new(string name, ovm_component parent = null);
105       super.new(name, parent);
106     endfunction
```

```
107
108      function void build();
109        p = new("producer", this);
110        c = new("consumer", this);
111        f = new("fifo", this);
112      endfunction
113
114      function void connect();
115        p.put_port.connect(f.blocking_put_export);
116        c.get_port.connect(f.blocking_get_export);
117      endfunction
118
119    endclass
file:  03_tlm/05_fifo/fifo.sv
```

The connect() function makes the association between the ports on the producer and consumer and the corresponding exports on the fifo. The run() task is responsible for controlling testbench execution. In this simple example, we let the testbench run for 100 ns and then terminate.

3.6   Forming a Transaction-Level Connection

To form a transaction-level connection, you must specify three elements: the control flow, the data flow, and the transaction data type. Declaring a connection as a port or export identifies the control flow — control flows from ports to exports. That is, a port initiates activity and an export responds to it. The interface identifies the data flow. A put interface indicates that data flows from the initiator (port side) to the target (export side), a get interface indicates that data flows from the target to the initiator, and a transport or request-response interface indicates a bidirectional data flow.

We declare put_port as a port, so we know the device in which this port is declared is an initiator. The interface type is tlm_nonblocking_put_if<>, which is one of the put interfaces defined in the TLM library. This port is an egress for data objects. Finally, the data type of the object being sent is trans.

In SystemVerilog using OVM, port and export declarations capture these three elements. Here is an example:

```
ovm_nonblocking_put_port #(trans) put_port;
```

The suffix of the object type is _port, indicating this is a port object. Exports use the suffix _export. The interface type is identified by the name between the ovm_ prefix and the _port or _export suffix. In this case, that name is nonblocking_put, which refers to tlm_nonblocking_put_if.

We have seen a producer and a consumer, each of which uses blocking tasks to send and retrieve transactions. The blocking tasks reside in the FIFO, an object that serves as an intermediary between the two components, otherwise known as a channel. The channel transfers data between the two components, and it serves as a synchronizing agent.

Putting a FIFO between two components to buffer and synchronize transfers is a common idiom in TLM. We will see this idiom frequently in the transaction-level testbenches we build using the OVM.

3.7   Summary

Put, get, and transport are fundamental means for synchronizing parallel processes and for communicating transaction-level information between those processes. These ideas are used extensively in the OVM to build transaction-level testbenches. Section 3.5 illustrated transaction-level communication using OVM facilities. In the next chapter, we will delve deeper into the OVM to show how to build arbitrary hierarchies of class-based verification components connected with transaction-level interfaces.

4

OVM Mechanics

The OVM library provides many facilities for constructing testbenches. In this chapter we will take a first look at the essential ones that you will use in almost all of your testbenches.

4.1　Components and Hierarchy

The primary structure for building testbench elements is the *component*. A component in OVM is analogous to a module in Verilog. An OVM component is constructed from a class, which gives it different characteristics than a Verilog module and has different usage implications. Amongst the different characteristics is that classes are created at run time, not at elaboration time as modules are. Therefore, OVM is responsible for creating the component instances and assembling them into hierarchies.

Figure 4-1 illustrates a simple hierarchy of components. Following, we will show how to build this hierarchy using the OVM facilities for creating components and composing them into hierarchies:

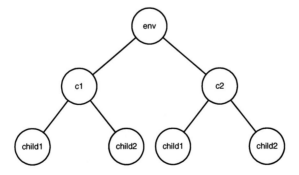

M. Glasser, *Open Verification Methodology Cookbook*, DOI: 10.1007/978-1-4419-0968-8_4,
© Mentor Graphics Corporation, 2009

Figure 4-1 A Simple Hierarchy of Components

The top-most node, env, is the *root*. The root is distinguished by the fact that it has no parent. All other nodes have exactly one parent. Each node has a name. The location in the hierarchy of each node can be identified by a unique *full_name* (path), which is constructed by stringing together the names of all the nodes between the root and the node in question, separating them with a hierarchy separator, dot (.). For example, the path to the component that is the second child of c2 is top.c2.child2.

A component in OVM is a class derived from ovm_component. The simplest components are *leaves*, those that have no children.

```
57    class child extends ovm_component;
58
59       function new(string name, ovm_component parent);
60          super.new(name, parent);
61       endfunction
62
63    endclass
file: 04_OVM_mechanics/01_hierarchy/top.sv
```

The constructor has two parameters, the name of the component and a pointer to its parent. The name is a simple name, not a hierarchical path. The parent provides a place to hook our new component into the hierarchy. A child's fully qualified path name is created by concatenating the child's name to the parent's full path name, separated by a dot (.). The OVM provides methods for retrieving both the name and fully qualified path of a component:

```
string get_name();

string get_full_name();
```

Subordinate components are instantiated in the build() function which is called during the build phase (phases are explained later in this chapter). Instantiating a component involves calling new() to allocate memory for it and passing the appropriate arguments into the constructor. In component, shown below, we instantiate two subordinate components, child1 and child2.

```
71    class component extends ovm_component;
72
73       child child1;
74       child child2;
75
```

```
76        function new(string name, ovm_component parent);
77            super.new(name, parent);
78        endfunction
79
80        function void build();
81            child1 = new("child1", this);
82            child2 = new("child2", this);
83        endfunction
84
85     endclass
file:  04_OVM_mechanics/01_hierarchy/top.sv
```

Like component, env also instantiates two subordinate components, c1 and c2. The entire hierarchy is rooted at a module called top in our design.

```
131    module top;
132
133      env e;
134
135      initial begin
136          e = new("env");
137          run_test();
138      end
139
140    endmodule
file:  04_OVM_mechanics/01_hierarchy/top.sv
```

The call to new() instantiates the top-level environment. run_test() starts execution of the testbench.

In SystemVerilog, modules, interfaces, and program blocks are created during elaboration, while classes are created after elaboration, at run time. So, to create a hierarchy of classes, we must have an interface, module, or program that contains an initial block that starts off the process of building a class-based component hierarchy. Interfaces are intended to serve as a medium of communication between two modules and are not well suited for serving as the root of a class-based hierarchy. Either program blocks or modules can be used to hold the root. For our simple hierarchy, it doesn't matter. Later, when we connect a class-based component to module-based hardware, we'll see that using a module is preferable to program blocks.

4.1.1 Traversing the Hierarchy

We can explore the data structures used to implement the component hierarchy with some methods provided in ovm_component. The children of a component are stored in an associative array. This array is not directly

accessible, but it can be accessed through a hierarchy API. This API is similar
to the built-in methods SystemVerilog provides for associative arrays.

```
int get_first_child(ref string name);

int get_next_child(ref string name);

ovm_component get_child(string name);

int get_num_children();
```

get_first_child() and get_next_child() work together to iterate over
the set of children contained in a component. get_first_child() retrieves
the name of the first child in the list. It returns the name as a reference
argument. get_next_child() returns the name of the next child in the list. It
returns 1 if there is a next child name to return or 0 if the end of the list has
been reached. get_child() transforms the name into a component reference.

Using these functions, we can traverse the component hierarchy.

```
73        function void depth_first(ovm_component node,
74                                  int unsigned level = 0);
75
76          string name;
77
78          if(node == null)
79            return;
80
81          visit(node, level);
82
83          if(node.get_first_child(name))
84            do begin
85              depth_first(node.get_child(name), level+1);
86            end while(node.get_next_child(name));
87
88        endfunction
file: 04_OVM_mechanics/utils/traverse.svh
```

This function will perform a depth-first traversal of the hierarchy, calling
visit() at each node. We use get_first_child() and get_next_child()
to iterate through the list of each of the children in each node. For each
iteration we call depth_first() recursively. For our small design, the result
is this:

```
+ env
|  + env.c1
|  |   env.c1.child1
|  |   env.c1.child2
```

```
|    + env.c2
|    |   env.c2.child1
|    |   env.c2.child2
```

The `visit()` function uses the node depth and whether or not it is a leaf node to print a line for each node.

4.1.2   Singleton Top

Components that don't have a parent (that is, the parent argument in the constructor is `null`) are called *orphans*. In OVM, you can create as many components without a parent as you like. However, there is no such thing as a true orphan. Any component whose parent is `null` is assigned a built-in parent called `ovm_top`. `ovm_top` is a singleton instance of `ovm_root`. It is the parent of all components that don't otherwise have a parent. In fact, `env` in our previous example is a child of `ovm_top`. Since it has no parent, it is automatically given `ovm_top` as its parent.

A *singleton* is a well-known, design object-oriented pattern characterized by a private (local) or protected constructor and a static get function that returns the same pointer no matter how many times it's called. This means it is only possible for one instance to exist, and that instance comes into existence automatically when `get()` is called. `ovm_top` contains a handle to the singleton instance of `ovm_root`. It is statically initialized by calling `ovm_root::get()`. You can call `ovm_root::get()` any time, but there is no need since `ovm_top` is provided as a convenience.

There are a number of beneficial consequences of having a singleton top-level component. One is that you can reach any component from `ovm_top`. If you run the hierarchy traversal algorithm on `ovm_top`, you will reach *every* component in the system. Another consequence is that any component, including ports, exports, and channels, that is instantiated inside a module is reachable from `ovm_top`. If you want to modify the report handlers in all components, for example, you can do so by calling one of the hierarchical reporting functions in `ovm_top`. `ovm_top` contains all the mechanisms for phasing, which is explained later in this chapter.

4.2   Connectivity

Components are connected to each other through TLM ports and exports. Ports and exports provide a means for components, or more accurately, processes in components, to synchronize and communicate with each other. Ports and exports are objects that form a binding point to enable intercomponent communication. As discussed in the previous chapter, exports provide functions and tasks that can be called by ports.

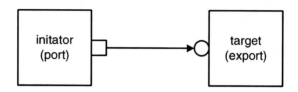

Figure 4-2 Connecting an Initiator to a Target

The connect method on ports and exports is used to bind the two together.

```
initiator_port.connect(target.export)
```

This method creates an association, or a *binding*, between the port and export so that the port can now call tasks and functions on the export. For the connection to be made successfully, the types of the port and export must match. That is, the interface types must be the same, and the type of the object being transferred in the interface must be the same.

4.2.1   Connecting across the Hierarchy

Similar to pins in an RTL design, we need to connect to TLM ports across hierarchical boundaries. Figure 4-3 uses a simple design to illustrate how to make these connections. This design contains a source component with two ports that ultimately connect to two exports, one on each of two sink components. To connect between these components, we must extend the ports and exports to the next level of hierarchy above.

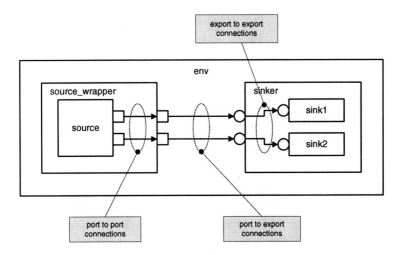

Figure 4-3 Connecting Ports and Exports through the Hierarchy

The component source contains two ports, first_put_port and second_put_port. These are instantiated in the build function.

```
65     class source extends ovm_component;
66
67        ovm_put_port #(trans_t) first_put_port;
68        ovm_put_port #(trans_t) second_put_port;
69
70        function new(string name, ovm_component parent);
71           super.new(name, parent);
72        endfunction
73
74        function void build();
75           first_put_port = new("first_put_port", this);
76           second_put_port = new("second_put_port", this);
77        endfunction
file: 04_OVM_mechanics/02_connectivity/top.sv
       . . .
```

Similarly, the sink component instantiates an export and instantiates it in the build function. The export is connected to an internal channel, fifo, from which the component can retrieve objects during run time.

```
126    class sink extends ovm_component;
127
128       ovm_put_export #(trans_t) put_export;
129       local tlm_fifo #(trans_t) fifo;
130
131       function new(string name, ovm_component parent);
```

```
132          super.new(name, parent);
133      endfunction
134
135      function void build();
136        put_export = new("put_export", this);
137        fifo = new("fifo", this);
138      endfunction
139
140      function void connect();
141        put_export.connect(fifo.put_export);
142      endfunction
file: 04_OVM_mechanics/02_connectivity/top.sv
     . . .
```

source_wrapper must create a connection between the internal source component and its outer boundary. It makes this connection by instantiating its own ports that have the same type as the type of the lower-level ports, in this case, those that belong to source.

```
98    class source_wrapper extends ovm_component;
99
100     source s;
101     ovm_put_port #(trans_t) put_port1;
102     ovm_put_port #(trans_t) put_port2;
103
104     function new(string name, ovm_component parent);
105       super.new(name, parent);
106     endfunction
107
108     function void build();
109       s = new("source", this);
110       put_port1 = new("put_port1", this);
111       put_port2 = new("put_port2", this);
112     endfunction
113
114     function void connect();
115       s.first_put_port.connect(put_port1);
116       s.second_put_port.connect(put_port2);
117     endfunction
118
119   endclass
file: 04_OVM_mechanics/02_connectivity/top.sv
```

After the ports in source_wrapper are instantiated, they are then connected to the ports in the lower-level source component via the connect method on the ports. Making exports visible to a higher level of hierarchy is done in much the same way as we see in sinker.

```
160   class sinker extends ovm_component;
```

```
161
162     ovm_put_export #(trans_t) first_put_export;
163     ovm_put_export #(trans_t) second_put_export;
164
165     sink sink1;
166     sink sink2;
167
168     function new(string name, ovm_component parent);
169        super.new(name, parent);
170     endfunction
171
172     function void build();
173        sink1 = new("sink1", this);
174        sink2 = new("sink2", this);
175        first_put_export = new("first_put_export", this);
176        second_put_export = new("second_put_export", this);
177     endfunction
178
179     function void connect();
180        first_put_export.connect(sink1.put_export);
181        second_put_export.connect(sink2.put_export);
182     endfunction
183
184   endclass
file: 04_OVM_mechanics/02_connectivity/top.sv
```

The two lower-level sink components and the exports are instantiated in the usual way. They are then connected using the connect method on the exports. Now we create a port-export connection between source_wrapper and sinker, also using the connect function.

```
192   class env extends ovm_component;
193
194     sinker s;
195     source_wrapper sw;
196
197     function new(string name, ovm_component parent = null);
198        super.new(name, parent);
199     endfunction
200
201     function void build();
202        s = new("sinker", this);
203        sw = new("source_wrapper", this);
204     endfunction
205
206     function void connect();
207        sw.put_port1.connect(s.first_put_export);
208        sw.put_port2.connect(s.second_put_export);
209     endfunction
210
211     task run;
```

```
212    global_stop_request();
213        endtask
214
215    endclass
file: 04_OVM_mechanics/02_connectivity/top.sv
```

For new users, it can often be confusing to determine which port or export object they should connect, and which object is the argument. You can easily figure it out by following the control flow through the system. The general rule is that the calling port invokes connect() using the called port or export as the argument. Figure 4-4 shows the flow of control through our hierarchical system.

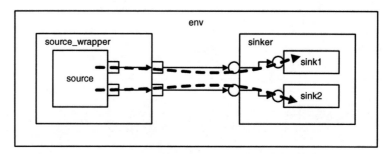

Figure 4-4 Control Flow through Ports and Exports

Ports are the site of the invocation and exports are the site of the invoked function or task. You can think of ports as calling exports. So, in env, we call connect on the put_ports supplying the put_exports as arguments. For port-to-port and export-to-export hierarchical connections, the calling order is a little less obvious. Since the call is made on the port side, you can think of the the lowest-level port in the hierarchy as calling interface methods in the upper-level port. Similarly, since exports are the site of the call, you can think of the upper-level export as calling into the lower-level export. The table below summarizes the possible connection types:

connection type	connection syntax
port-to-export	`port.connect(export);`
port-to-port	`child.port.connect(port);`
export-to-export	`export.connect(child.export);`

4.2.2   Note to AVM Users

In AVM-3.0, connections were made in a similar fashion using the connect call on ports and exports. In addition, the export-to-export, port-to-export, and port-to-port calls were made in different phases, export_connections(), connect(), and import_connections(), respectively. In OVM, the order in which the connect calls are made is no longer important; they can be made in any order. We recommend that you put them in the connect() phase (not to be confused with the connect() method on ports and exports).

OVM supports delayed binding, a feature where calls to connect() only make a note that a connection is to be made. Later, just before end_of_elaboration, the notes are reconciled and the connections made. This enables a cleaner use model and is much more forgiving of understandable errors where connect() calls were made in the wrong order.

4.3   Phases

Traditional Verilog modules rely on the simulator to elaborate the complete design and kick off its execution. Since OVM components are classes, they are instantiated and connected, and their execution is initiated outside of the Verilog elaborator. Components come into existence by calling class constructor new(), which allocates memory and performs initializations. Rather than the Verilog run-time engine managing instantiation, elaboration, and execution of class-based components, component functionality is broken into phases, and the OVM phase controller manages their execution.

Each phase is represented in the component as a virtual method (task or function) with a trivial default implementation. These *phase callbacks* are implemented by the component developer, who supplies appropriate functionality. The phase controller ensures that the phases are executed in the proper order. The set of predefined phases is shown in the following table:

phase name	function/ task	order
new	function	top-down
build	function	top-down
connect	function	bottom-up
end_of_elaboration	function	bottom-up

phase name	function/task	order
start_of_simulation	function	bottom-up
run	task	bottom-up
extract	function	bottom-up
check	function	bottom-up
report	function	bottom-up

Each phase has a specific purpose. Component builders must take care to ensure that the functionality implemented in each phase callback is appropriate to the phase definition.

- **new** is not technically a phase, in that it's not managed by the phase controller. However, for each component, the constructor must execute and complete in order to bring the component into existence. Therefore, `new()` must run before `build()` or any other subsequent phases can execute.
- **build** is the place where new components, ports, and exports are instantiated and configured. This is also the recommended place for calling `set_config_*` and `get_config_*` (see Section4.4).
- **connect** is where components, ports, and exports created in `build()` are connected.
- **end_of_elaboration** is where you can make configuration changes, knowing that elaboration is complete. That is, you can assume that all components are built and connected.
- **start_of_simulation** executes just before time 0.
- **run** is the only pre-defined *task* phase. All of the run tasks are forked to run in parallel. Each run task continues until its locus of control passes the endtask statement or it is explicitly shut down. Later in this chapter, we will discuss how to shut down testbenches.
- **extract** is intended for collecting information relating to coverage or other information about how to answer the testbench questions.
- **check** is where any correctness checking or validation of extracted data is done.
- **report** is where final reports are produced.

The simple example below uses `ovm_report_info()` calls to illustrate the order in which phases are executed.

```
38    class sub_component extends ovm_component;
39
40       function new(string name, ovm_component parent);
41          super.new(name, parent);
42       endfunction
43
44       function void build();
45          ovm_report_info("build", "");
46       endfunction
47
48       function void connect();
49          ovm_report_info("connect", "");
50       endfunction
51
52       function void end_of_elaboration();
53          ovm_report_info("end_of_elaboration", "");
54       endfunction
55
56       function void start_of_simulation();
57          ovm_report_info("start_of_simulation", "");
58       endfunction
59
60       task run();
61          ovm_report_info("run", "");
62       endtask
63
64       function void extract();
65          ovm_report_info("extract", "");
66       endfunction
67
68       function void check();
69          ovm_report_info("check", "");
70       endfunction
71
72       function void report();
73          ovm_report_info("report", "");
74       endfunction
75
76    endclass
file: 04_OVM_mechanics/03_phases/top.sv
```

In a top-level component, we create two instantiations of component, each of
which in turn instantiates two sub_components. The sub_components are
essentially the same as component; each phase callback simply prints a line
identifying the phase. When executed, you get the following result:

```
OVM_INFO @ 0 [RNTST] Running test ...
OVM_INFO @ 0: env.c1 [build]
OVM_INFO @ 0: env.c1.s1 [build]
OVM_INFO @ 0: env.c1.s2 [build]
OVM_INFO @ 0: env.c2 [build]
```

```
OVM_INFO @ 0: env.c2.s1 [build]
OVM_INFO @ 0: env.c2.s2 [build]
OVM_INFO @ 0: env.c1.s1 [connect]
OVM_INFO @ 0: env.c1.s2 [connect]
OVM_INFO @ 0: env.c1 [connect]
OVM_INFO @ 0: env.c2.s1 [connect]
OVM_INFO @ 0: env.c2.s2 [connect]
OVM_INFO @ 0: env.c2 [connect]
OVM_INFO @ 0: env.c1.s1 [end_of_elaboration]
OVM_INFO @ 0: env.c1.s2 [end_of_elaboration]
OVM_INFO @ 0: env.c1 [end_of_elaboration]
OVM_INFO @ 0: env.c2.s1 [end_of_elaboration]
OVM_INFO @ 0: env.c2.s2 [end_of_elaboration]
OVM_INFO @ 0: env.c2 [end_of_elaboration]
OVM_INFO @ 0: env.c1.s1 [start_of_simulation]
OVM_INFO @ 0: env.c1.s2 [start_of_simulation]
OVM_INFO @ 0: env.c1 [start_of_simulation]
OVM_INFO @ 0: env.c2.s1 [start_of_simulation]
OVM_INFO @ 0: env.c2.s2 [start_of_simulation]
OVM_INFO @ 0: env.c2 [start_of_simulation]
OVM_INFO @ 0: env.c2 [run]
OVM_INFO @ 0: env.c2.s2 [run]
OVM_INFO @ 0: env.c2.s1 [run]
OVM_INFO @ 0: env.c1 [run]
OVM_INFO @ 0: env.c1.s2 [run]
OVM_INFO @ 0: env.c1.s1 [run]
OVM_INFO @ 1: env.c1.s1 [extract]
OVM_INFO @ 1: env.c1.s2 [extract]
OVM_INFO @ 1: env.c1 [extract]
OVM_INFO @ 1: env.c2.s1 [extract]
OVM_INFO @ 1: env.c2.s2 [extract]
OVM_INFO @ 1: env.c2 [extract]
OVM_INFO @ 1: env.c1.s1 [check]
OVM_INFO @ 1: env.c1.s2 [check]
OVM_INFO @ 1: env.c1 [check]
OVM_INFO @ 1: env.c2.s1 [check]
OVM_INFO @ 1: env.c2.s2 [check]
OVM_INFO @ 1: env.c2 [check]
OVM_INFO @ 1: env.c1.s1 [report]
OVM_INFO @ 1: env.c1.s2 [report]
OVM_INFO @ 1: env.c1 [report]
OVM_INFO @ 1: env.c2.s1 [report]
OVM_INFO @ 1: env.c2.s2 [report]
OVM_INFO @ 1: env.c2 [report]
```

You can see that build() runs top-down and the rest of the phases run bottom-up. You can also see that each phase completes in all components before the next phase begins. Thus, in connect(), for example, you can rely on the fact that build() has completed in all components. You will also notice that time advances after the run phase. In our example the run() task is trivial; it simply delays one time unit (#1).

run_test(), mentioned previously in Section 4.1, initiates executions of the phases. It starts running the phases in order and controls the machinery for making sure each phase is complete before the next one begins.

4.4   Config

To increase reusability of components, it's desirable to sprinkle them liberally with parameters that can be externally configured. The config facility provides a means to do just this. It is based on a database of name-value pairs called *configuration items*[1] that is organized hierarchically. Each component contains a configuration table of configuration items and, since components are arranged in a tree, each element in the database can be uniquely located by the location of the component and the name of the configuration item.

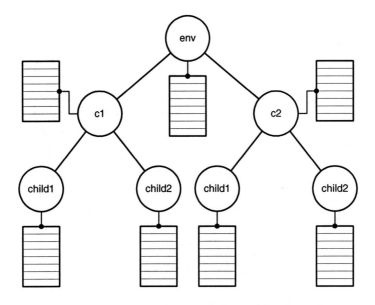

Figure 4-5 Each Component Has a Database of Configuration Items

---

1. We use the term *configuration item* instead of *parameter* to avoid confusion with other uses of the term parameter in SystemVerilog.

The `ovm_component` class contains two sets of methods for putting configuration items into the database and for retrieving them later. These are `set_config_*` and `get_config_*`. The table below shows both sets.

Configuration Database Access Functions
`set_config_int(string name, string field_name, int value)`
`set_config_string(string name, string field_name,` `                   string value)`
`set_config_object(string name, string field_name,` `                   ovm_object object, bit clone);`
`get_config_int(string field_name, inout int value);`
`get_config_string(string field_name, inout string value);`

`get_config_object(string field_name,`
`                   inout ovm_object object,`
`                   input bit clone);`

The `set_config_*` functions place an item in the configuration database in the current component, that is, in the component instance in which the function is called. These functions each take three arguments, `name`, `field_name`, and `value`. The argument `name` is a path name that represents the scope of the components that are to accept this configuration item. `name` is used in `get_config_*` to locate items in the configuration database. `field_name` is the name of the field and must be unique within the current configuration database. `value` is the value part of the name-value pair and its type can be `string`, `int`, or `ovm_object`, depending on which function is being called. In addition, `set_config_object` takes a clone argument to indicate whether the object being passed in as the value should be cloned before it is put into the configuration database.

The `get_config_*` functions retrieve items from the configuration database. These functions take only two arguments, a field name and an inout variable that contains the value of the item located. They also return a bit to indicate whether the requested item was successfully located. The `get_config_*` functions do not take a path name argument like their `set_config_*` counterparts because they use the path of the current component as the point of reference to locate configuration items. They are designed to inquire as to the value of a configuration item for the current context, that is, the component in which the `get_config_*` function is called.

The search algorithm for retrieving configuration items uses the path name of the component requesting a configuration item and the path name inserted in each item. It starts by looking up the config item in the database in the top-most component (the singleton top) by field_name. If such an item exists, it then asks if the path name specified in the item matches the path name of the component. If an item with the specified field_name is not located or the path names do not match, then the search proceeds with the child component. This process continues until a match is made or the search reaches the component where the search originated.

The path name in each configuration item can be a regular expression. So we use a regular expression matching algorithm to match the requested component path name and configuration item path name. The effect is to match hierarchical scopes.

As an example, consider the simple hierarchy in Figure 4-5. Let's say that in env::build() we issue two set_config_* calls:

```
112      function void build();
113          c1 = new("c1", this);
114          c2 = new("c2", this);
115
116          set_config_int("c2.*", "i", 42);
117          set_config_int("*", "t", 19);
118      endfunction
file: 04_OVM_mechanics/04_config/top.sv
```

This will cause two configuration items to be entered into the database in env. Notice the asterisk (*) in the path names. Path names in calls to set_config_* are regular expressions, and wild card characters are used to specify multiple scopes over which the configuration item applies. For item i, c2.* indicates that in any scope below c2 in the component hierarchy, i will take the specified value. In this case, the specified value is 42. If you leave off the asterisk, then the configuration item applies only to c2 and not to any of its children.

The state of the configuration databases for each component in the hierarchy after the set_config_* calls are made is shown in the Figure 4-6,

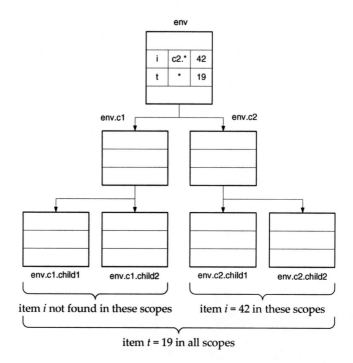

Figure 4-6 Hierarchy of Configuration Databases

Now, let's say that in top.c1.child1 we issue the call:

```
int i;
...
get_config_int("i", i)
```

The search asks the question What is the configuration value for i in the hierarchical scope top.c1.child1? To answer this question, the configuration database in env is searched first. The entry for i there says that the value of i in scopes matching env.c2.* is 42. However, the component from which the request was issued is in the c1 sub-hierarchy. Therefore, there is no match, and the get_config_int() call returns a failure status. A request in any component that is a child of c2 would successfully complete and return a value of 42.

Below is the code for the build() function of the child components. This is where these components look up configuration values for i and t.

```
60        function void build();
61
```

```
62          string msg;
63
64          if(!get_config_int("t", t)) begin
65              $sformat(msg, "no value for t found in config
database, using default value of %0d", t);
66              ovm_report_warning("build", msg);
67          end
68
69          if(!get_config_int("i", i)) begin
70              $sformat(msg, "no value for i found in config
database, using default value of %0d", i);
71              ovm_report_warning("build", msg);
72          end
73
74      endfunction
file: 04_OVM_mechanics/04_config/top.sv
```

The following sample shows the printed output from running this design:

```
# OVM_INFO @ 0 [RNTST] Running test ...
# OVM_WARNING @ 0: env.c1.child1 [build] no value for i found in
config database, using default value of 91
# OVM_WARNING @ 0: env.c1.child2 [build] no value for i found in
config database, using default value of 91
```

The request for configuration item t was successful in all contexts since the set_config_int call established that t is available in all contexts. Two of the requests for configuration item i succeeded, and two failed. This outcome is because we limited the availability of i to only the components at or below env.c2. The components at or below c1 cannot see the configuration item i because of the way we have established the configuration database.

4.4.1   Configuration and Phasing

Now that we know the set of calls for putting items into the configuration database and retrieving them, the next task is to effectively apply those functions to configure components. Configuration can be used to alter the behavior or the structure of a testbench. Typically, the behavior modes and structure are determined when the testbench begins, so it is most useful to establish the configuration settings in one of the early phases, such as new, build, or connect.

From the table of phases on page 79, you can see that new and build are top-down phases, while all the remaining phases are bottom-up. So, if you want to set a configuration item in the database at a higher-level context to be picked up by a lower-level one, you must call set_config_* in either the new or build phase. The phases are executed discretely, meaning that each phase runs to completion before the next phase begins. You can set configuration items in either the new or build phase and retrieve them for use in setting behavior modes or modifying topology in the build phase. In your build function, first make `get_config_*` calls to retrieve items from higher levels of hierarchy to control the configuration of the current level. Next, add `set_config_*` calls to put configuration items into the database for use by components at lower levels of hierarchy. Finally, using the appropriate configuration items, instantiate components. It is important to call `get_config_*` first, because the information specified may affect the values that are set for lower levels of the hierarchy.

For example, configuring topology involves setting the topology parameters in the top-level environment and then applying those parameters in various components that are below the top-level environment in the hierarchy. Our example has a bus that can have any number of masters or slaves. The number of masters and slaves is set in the top-level environment. The bus model picks up this configuration information and uses it to construct the bus. In the build function of the top-level environment, we instantiate the bus model and configure it with the number of masters and slaves we want it to have.

```
129      function void build();
130        set_config_int("bus", "masters", 4);
131        set_config_int("bus", "slaves", 8);
132        b = new("bus", this);
133      endfunction
file: 04_OVM_mechanics/05_config_topo/top.sv
```

The bus model is constructed so that the number of masters and slaves is not fixed. Instead, those numbers come from the configuration system.

```
90       function void build();
91
92         int unsigned i;
93
94         if(!get_config_int("masters", masters)) begin
95           $sformat(msg, "\"masters\" is not in the
configuration database, using default value of %0d", masters);
96           ovm_report_warning("build", msg);
97         end
98
99         for(i = 0; i < masters; i++) begin
100          $sformat(name, "master_%0d", i);
101          m = new(name, this);
102        end
103
```

```
104          if(!get_config_int("slaves", slaves)) begin
105            $sformat(msg, "\"slaves\" is not in the configuration
database, using default value of %0d", slaves);
106              ovm_report_warning("build", msg);
107          end
108
109          for(i = 0; i < slaves; i++) begin
110            $sformat(name, "slave_%0d", i);
111            s = new(name, this);
112          end
113
114       endfunction
file: 04_OVM_mechanics/05_config_topo/top.sv
```

In the build function for the bus model, the design retrieves the required configuration information using calls to get_config_int. In each case, the return value is checked to determine whether the requested config item was successfully retrieved. If not, a warning message issues, noting that the config item was not found and that the default value will be used. From a best-practices perspective, it is important to make sure that the return value is checked and a warning is issued if it indicates failure. Without that check, the fact that the default value is being used could go unnoticed. In some cases it may be acceptable to use the default; in other cases it is not acceptable. The person building the bus model may not know all the circumstances under which the model will be used. So it is important to do everything possible to make the model robust. Checking status returns and issuing messages as appropriate is one way to improve the robustness of a model.

In the loop where we build the masters, the reference to each new master is saved in the same variable, m. Each new master overwrites the previous one. We don't bother using an array to save all the component handles. In each iteration of the loop, we use $sformat to generate a unique name. The constructor, new(), calls super.new(), the constructor in the ovm_component base class that is responsible for inserting the newly created component into the parent's list of children components. There is no need to explicitly save the component handles because the parent component does that for us. The loop that creates the slaves is organized the same way.

4.5 Factory

The structure of a testbench is determined by the organization of the components into a hierarchy and the way these objects are connected. The behavior of the testbench is determined by the procedural code in the phase callbacks—build, connect, run, and so forth. There are times when it is desirable to modify the behavior or part of the structure externally, that is, at run time, without touching the testbench code. For example, to inject errors into a system, you may want to replace the normal driver with an error driver, one that intentionally injects errors. Instead of re-coding the environment to use a different driver, you can use the factory to do the substitution automatically.

The *factory* provides a means for substituting one object for another without having to use your text editor to modify the testbench. Instead of creating the object using `new()`, you invoke a create function in the factory. The factory keeps a list of registered objects and, optionally, a set of overrides associated with each one. When you create an object using the factory, the list of overrides is consulted. If one is present, then the override object is returned. Otherwise, the registered object is returned.

The factory is an OVM data structure. It is global in scope, and only one instance exists (that is, it's a singleton). It serves as a *polymorphic constructor,* a single function that lets you build a variety of different objects. It provides a means for registering objects and for specifying overrides. Objects registered as overrides must be derived from the object they are overriding. To have a single function return multiple objects, each of those objects must be derived from a common base class.

An essential component of the factory is the wrapper, a class that wraps the object we wish to register with the factory. The factory data structure is a table of wrappers indexed by a key. The wrapper has a `create()` function that delegates to the constructor of the wrapped object.

Using the factory involves three steps: *registration, setting overrides,* and *creation.* In the first step, you register an object with the factory. In the second step, you add an override to a registered object. In the third step, you create an object with the factory that will return either the originally registered object or an override, depending on whether an override was registered for the requested object.

4.5.1   How the Factory Works

The term *factory* was coined for use in the software world by The Gang of Four in their book *Design Patterns: Elements of Reusable Object-Oriented Software.* In that book, they identified the pattern, which they call *abstract factory,* as an interface for creating families of related objects. They identified the pattern *factory method* as an interface for creating objects, but defer to subclasses for decisions about which object to create. The OVM factory is a combination of both of these creation patterns. It provides a means to create a family of objects, and it provides a means for delegating the decision as to exactly which object to create to the factory data structure.

The OVM factory is based around a data structure that maps requested types to override types. Essentially, the organization is an associative array of type handles whose key is also a type handle. When a type is registered with the factory, its override type is itself. So by default, when you request an object of that type, you get only that type. The factory also provides a means for replacing the overrides with other types so you can retrieve override types that are different than the registered type.

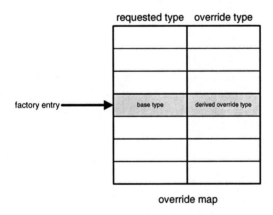

Figure 4-7 Factory Override Map Data Structure

The following example is a highly simplified toy factory that illustrates how the OVM factory works. The toy factory retains the essential structure of the OVM factory, but many details have been removed for the purposes of keeping the illustration clear. Our toy factory is implemented in four classes, two base classes and two derived classes, that do the real work. The two base classes are object_base and wrapper_base. All objects registered in the factory must be derived (ultimately) from object_base, while wrapper_base is the base class for the parameterized wrappers. factory is a singleton that contains the associative array of type handles that are instances of wrappers. Finally, wrappers are derived from wrapper_base and are parameterized classes that represent unique types.

For our toy factory, the base classes are trivial:

```
46      class object_base;
47          virtual function void print();
48              $display("object_base");
```

```
49        endfunction
50      endclass
file:  04_OVM_mechanics/07_toy_factory/top.sv

59      class wrapper_base;
60        virtual function object_base create_object();
61          return null;
62        endfunction
63      endclass
file:  04_OVM_mechanics/07_toy_factory/top.sv
```

object_base has a virtual print() function, which we use to verify the types of objects created by the factory. wrapper_base has the virtual function create(), the polymorphic constructor function that is used to create new objects.

factory is a singleton, meaning its constructor is local, and it contains a static reference to an instance of itself. The only way to create an instance of factory is to call factory::get(). factory contains an associative array that maps wrapper_base handles of requested types to wrapper_base handles of override types.

```
73      class factory;
74
75        static factory f;
76        wrapper_base override_map[wrapper_base];
77
78        local function new();
79        endfunction
80
81        static function factory get();
82          if(f == null)
83            f = new();
84          return f;
85        endfunction
86
87        function void register(wrapper_base w);
88          override_map[w] = w;
89        endfunction
90
91        function void set_override(wrapper_base requested_type,
92                                   wrapper_base override_type);
93          override_map[requested_type] = override_type;
94        endfunction
95
96        function object_base create(wrapper_base
97                                     requested_type);
98          object_base obj;
99          wrapper_base override_type =
```

```
100             override_map[requested_type];
101          obj = override_type.create_object();
102          return obj;
103       endfunction
104
105   endclass
file: 04_OVM_mechanics/07_toy_factory/top.sv
```

The register() method adds a new entry to the override map. Initially, upon registration, a type has no overrides. Therefore, we set the override map to map a type handle to itself. The set_override() method replaces the entry in the override map with a new override type. The create() method looks up the override for the requested type, delegates creation to the override type, and returns the newly created object.

The wrapper class is the most interesting class in our constellation of factory-related classes. Even though it's quite simple, it does most of the heavy lifting. It is the primary interface to the factory, and most of the operations you do with the factory, you do though the wrapper interface.

```
118   class wrapper #(type T=object_base) extends wrapper_base;
119
120      typedef wrapper#(T) this_type;
121
122      static this_type type_handle = get_type();
123
124      local function new();
125      endfunction
126
127      function object_base create_object();
128        T t = new();
129        return t;
130      endfunction
131
132      static function T create();
133        T obj;
134        factory f = factory::get();
135        assert($cast(obj, f.create(get_type())));
136        return obj;
137      endfunction
138
139      static function this_type get_type();
140        factory f;
141        if(type_handle == null) begin
142          type_handle = new();
143          f = factory::get();
144          f.register(type_handle);
145        end
146        return type_handle;
```

```
147        endfunction
148
149        static function void set_override(wrapper_base
150                                            override_type);
151          factory f = factory::get();
152          f.set_override(type_handle, override_type);
153        endfunction
154
155     endclass
file: 04_OVM_mechanics/07_toy_factory/top.sv
```

All of the functions in `wrapper#()` are static except for the constructor and `create_object()`. We can execute its static functions without concern for whether it has been explicitly instantiated. Since each wrapper specialization is a singleton, there can be no more than one instance of it. That means that the static member `type_handle` is unique and can be used as a proxy for the wrapped type (that is, the type supplied as a parameter that is used to specialize the class). Since the type handle is unique and most of the methods are static, we can treat the type handle more like a type than an object.

The type handle is initialized statically. This occurs in the following line:

```
static wrapper#(T) type_handle = get_type();
```

The function `get_type()` is called during static initialization, which not only creates an instance of the wrapper, but also registers it in the factory. To register a class with factory, you first specialize a wrapper with the type of the object you are wrapping. Use a typedef to specialize the wrapper, as shown in the following example:

```
typedef wrapper#(some_type) type_id;
```

This typedef creates a wrapper for type `some_type`, a type derived from `object_base`.

Figure 4-8 illustrates how to use our toy factory with some toy classes A, B, and C, which are derived from `family_base`.

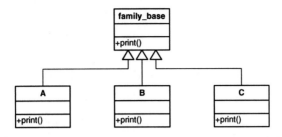

Figure 4-8 Family of Classes for the Toy Factory

To register the classes with the factory, each of them has a typedef of the wrapper parameterized with its own type. Below is class A. Classes B and C are similar. Each has a typedef that specializes the wrapper.

```
169    class A extends family_base;
170
171       typedef wrapper#(A) type_id;
172
173       virtual function void print();
174          $display("A");
175       endfunction
176    endclass
file: 04_OVM_mechanics/07_toy_factory/top.sv
```

The following is a short program that exercises the factory.

```
206    function void run();
207
208       f = factory::get();
209
210       h = family_base::type_id::create();
211       h.print();
212
213
family_base::type_id::set_override(B::type_id::get_type());
214
215       h = family_base::type_id::create();
216       h.print();
217
218    endfunction
file: 04_OVM_mechanics/07_toy_factory/top.sv
```

The code makes heavy use of the double-colon (::) scope operator. It is used to refer to static functions in factory and in wrapper#(). First, we get the singleton instance of the factory data structure, then we ask to create an instance of object `family_base`. `family_base::type_id::get_type()` is the static function inside the wrapper specialization for `family_base`. We verify that an instance of `family_base` is created by calling `print()`. Next, we set an override of B for `family_base`. Again, we create an instance of `family_base`. This time, since an override is now in place, instead of getting an instance of `family_base`, we get an instance of B.

Our toy factory does not contain all of the functionality of the factory implemented in OVM. The OVM factory provides a mapping from strings (names) to type handles. It allows override chaining; whereas, our toy factory does not. For example, if B overrides A, and C overrides B, when you ask for an instance of A, you will get an instance of C. The OVM factory supports two primary base classes, `ovm_object` and `ovm_component`, for registered objects, and provides `create()` methods for both; whereas, the toy factory has only one primary class, `object_base`.

4.5.2   The OVM Factory API

In this section, we will look more closely at the OVM factory API. It has two parts, the type-based factory and the string-based factory. In the string-based factory, requested types are identified by string names. In the type-based factory, types are identified by type handles. A single type can be registered both ways. The methods for performing the three steps (registration, setting overrides, and creation) are slightly different in each way. First, we'll look at the type-based factory. Here is a component called `driver` that registers itself with the type-based factory.

```
49    class driver extends ovm_component;
50
51      typedef ovm_component_registry#(driver) type_id;
52
53      static function type_id get_type();
54        return type_id::get();
55      endfunction
56
57      function string get_type_name();
58        return "driver";
59      endfunction
60
61      function new(string name, ovm_component parent);
62        super.new(name, parent);
63      endfunction
64
65    endclass
file: 04_OVM_mechanics/06_factory/top.sv
```

There are two parts to registration, supplying the typedef of
`ovm_component_registry#()` and supplying the static function `get_type()`.
The typedef creates a specialization of `ovm_component_registry` using the
component type `driver` as the type parameter. The
`ovm_component_registry#()` class has a static initializer that does the
registration with the factory data structure. So, creating the specialization
using a typedef causes the class identified by the parameter, `driver` in this
case, to be registered into the factory.

Setting an override is a simple matter of calling the `set_override` function
and supplying the override type as an argument, as shown below:

```
105
driver::type_id::set_type_override(error_driver::get_type());
file: 04_OVM_mechanics/06_factory/top.sv
```

This string looks like quite a mouthful, but it really is quite simple. Let's
deconstruct the statement to fully understand what it means.

`driver` – the requested type.

`driver::type_id` – the type of the specialized wrapper.

`driver::type_id::set_type_override` – the set override function
in the specialized wrapper. This is a static function, which is why you
need the `::` scope operator to refer to it.

`error_driver` – the override type.

`error_driver::get_type()` – the static function that returns the
type handle for `error_driver`.

To create an instance of a class using the factory, we call the create method in
the factory, as shown below:

```
102        d1 = driver::type_id::create("d1", this);
file: 04_OVM_mechanics/06_factory/top.sv
```

The `::` syntax works the same as it does in the previous example—`driver`
refers to the requested type, `driver::type_id` refers to the type of the
specialized wrappers, and `driver::type_id::create` refers to the create
function in the specialized wrapper. This statement creates an instance of
`driver`. The difference between calling `create()` and calling `new()` is that

create() will consult the factory to see if there are any overrides. If so, the create() will actually return an object of the override type.

Now let's look at the string-based factory. The registration mechanism for the string-based factory relies on a typedef just like the type-based factory, as shown in the line below:

```
48      typedef ovm_component_registry#(driver, "driver")
type_id;
```

The only difference is the addition of the second parameter of the parameterized wrapper. It identifies the name of the type, in this case driver. When ovm_component_registry#() is specialized with two parameters, a type and a name, the wrapper is registered both with the type-based factory and the string-based factory. To set an override using the type name, call the factory API directly rather than use the wrapper API, as shown below:

```
106     factory.set_type_override_by_name("driver",
107                                       "error_driver");
```

This statement simply says, when requested to create an object whose type name is driver, return a type whose name is error_driver instead. The greatest difference between using the string-based and type-based factory is in how objects are created. In the type-based API, you access the factory through the wrapper. In the string-based API, you invoke the factory directly. The return type of ovm_factory::create_component_by_name() is ovm_component. Contrast this to the return type of ovm_component_registry#(T)::create, which is T. To access the intended type of the returned object you will have to downcast it, as shown below:

```
99      assert($cast(d1,
100             factory.create_component_by_name("driver",
101                                              "",
102                                              "d1",
103                                              this)));
```

The call to factory.create_component_by_name() returns an object of type ovm_component. The $cast downcasts the returned object to the type of d1, which is driver. Since the type name argument to create_component_by_name() is a string, there is no compile-time type checking to ensure that the type of the returned object can be cast to the required type. So it is important to check the return code of $cast to determine whether the cast succeeded.

The factory string-based API includes `create_object_by_name()`. It is used to create objects derived from `ovm_object`. You must call $cast to downcast the created object for the same reason you call cast on components created by the string-based factory API.

4.5.3   String-Based or Type-Based?

The type-based and string-based factory APIs each have their pros and cons, but generally, we recommend you use the type-based factory. It is far more robust, being removed from errors in string names.

Sometimes there is no other choice; only the string-based factory will do. An important example of this is when you want to specify a test name from the command line. `run_test()` takes an optional string argument, `test_name`. Also, this task looks at the command line argument `OVM_TESTNAME`. If a test name is supplied either through the command line or through the argument list, `run_test()` invokes the string-based factory to instantiate the test object. More about using command line arguments and the factory to choose a test is in Section 7.5.

The string-based factory suffers from two major drawbacks. One, just mentioned, is that it's easy to mis-type a type name when you are writing code. This can result in a broken testbench because an object is not located, or in a subtle bug where the *wrong* object is instantiated. The second drawback is that it's difficult, if not impossible, to represent parameterized classes using string names. For example, consider the parameterized class `my_class`.

```
class my_class #(type T=int) extends ovm_object;
    typedef my_class#(T) this_t;
    typedef ovm_object_registry#(this_t, "my_class#(T)") type_id;
endclass
```

We've registered it with the string-based factory using the name `my_class#(T)`. Seems logical. Now consider two specializations of that class.

```
typedef my_class#(A) C1;
typedef my_class#(B) C2;
```

There is no convenient way to register these with the factory using names that are unique to the specialization. In our example classes, `C1` and `C2` are each registered using the name `my_class #(T)`. To the factory data structure, it looks like you are trying to register two objects with the same name, which is an error. The remedy is to use the type-based factory API, which is not encumbered with strings.

```
class my_class #(type T=int) extends ovm_object;
  typedef my_class#(T), this_t;
  typedef ovm_object_registry#(this_t) type_id;
endclass
```

By leaving off the second argument in the typedef of `type_id`, we are telling the factory to register the object without a name, to only use the type handle as the lookup key. Now the specializations will each have their own unique type handle and will not be erroneously treated as the same object in the factory.

The drawback to using the type-based factory is that there is no way to look up an object by string. That's because with no second argument in the `ovm_object_registry` typedef, there is no name under which to file the object. When you first use the type-based factory, it may seem a bit disconcerting that no name is available. You will quickly discover that a string name is not really necessary. In the cases where a name is necessary, as when you are getting object names from user input, then the string-based factory is available.

4.6   Shutting Down the Testbench

The easiest way to shut down an OVM testbench is to call the global function `global_stop_request()`. This *requests* that the testbench shut down. If there is no reason not to shut down, then the testbench will terminate. `global_stop_request()` delegates to `ovm_top.stop_request()`. The two forms are semantically equivalent.

What reasons would there be to not allow a shutdown? Every component has a virtual task `stop()`. When you call `global_stop_request()`, this task is called for each component whose member `enable_stop_request` is set to 1. When all of the stop tasks have returned, then the testbench shuts down. The stop task can be used to clean things up, tell the DUT to shut down, serve as a shutdown objection, or anything else you'd like to do before completing the current phase. Since it is a task, `stop()` can consume time. A `stop()` task can disallow the shutdown by blocking. It can wait for some condition to be set or delay a fixed time. The `stop()` task services the shutdown request. When `stop()` returns, it allows the request to be granted.

The following example illustrates how the stop mechanism works. This example consists of two producers sending transactions to a consumer through a FIFO. Each producer runs independently of the other. We want to make sure that both producers finish their respective jobs. When one finishes, the other continues until it is done.

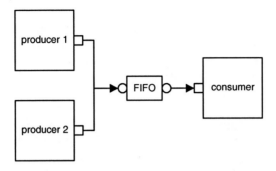

Figure 4-9 Two Producers and a Consumer

In the build() function for the top-level environment, in addition to instantiating the various components, we configure a different number of iterations in each producer. Since the number of iterations for each is different, one producer will finish before the other.

```
127      function void build();
128          set_config_int("producer1", "iterations", 5);
129          set_config_int("producer2", "iterations", 9);
130          p1 = new("producer1", this);
131          p2 = new("producer2", this);
132          c = new("consumer", this);
133          f = new("fifo", this);
134      endfunction
file: 04_OVM_mechanics/09_shutdown/top.sv
```

We want the testbench to shut down in an orderly fashion when all the required work is done, but we don't want it to shut down prematurely when the first producer completes. We use the stop mechanism to accomplish this objective. Each producer has a stop task that waits until done becomes 1.

```
77      task stop(string ph_name);
78          ovm_report_info("stop", "initating stop");
79          wait(done == 1);
80          ovm_report_info("stop", "shutting down...");
81      endtask
file: 04_OVM_mechanics/09_shutdown/top.sv
```

The ph_name argument contains the name of the phase in which stop() was called. Even though by default there is only a task-based phase, run(), it's possible to arbitrarily add more task-based phases (and function-based phases, too). The stop request mechanism works in all task-based phases, and a call to global_stop_request() operates in the current task-based phase. It causes stop() to be called. Since only one task named stop is possible in each component, the ph_name argument identifies the phase in which it was called. You can use that to modify the behavior of stop() based on the task-based phase name.

To use the stop request mechanism, we have to enable it by setting enable_stop_interrupt to 1. We do this in the component's constructor.

```
42        function new(string name, ovm_component p = null);
43          super.new(name,p);
44          iterations = 10; // default value
45          done = 0;
46          enable_stop_interrupt = 1;
47        endfunction
```

The main loop of the producer is straightforward. Each iteration of the loop generates a random integer and sends it through the put port to the consumer. In the following sample, when the loop completes, we set done to 1, which releases the stop tasks.

```
66        for(int i = 0; i < iterations; i++) begin
67          randval = $random % 100;
68          $sformat(s, "sending    %4d", randval);
69          ovm_report_info("producer", s);
70          put_port.put(randval);
71        end
72
73        done = 1;
file: 04_OVM_mechanics/09_shutdown/top.sv
```

The final part is the run task in the top-level environment:

```
142       task run();
143         ovm_report_info("run", "start");
144         global_stop_request();
145       endtask
file: 04_OVM_mechanics/09_shutdown/top.sv
```

Upon starting, the task immediately calls global_stop_request(), which causes the stop() tasks to be called in the producers (because enable_stop_interrupt is set to 1 in each). In turn, each producer blocks until its respective done flag is set. When the producer with the smallest number of iterations finishes, it triggers its local done flag and its stop task returns. However, because there are outstanding blocked stop tasks, the simulation continues. Only when all of the stop tasks complete will the simulation terminate.

4.6.1   Timeout

It's possible that a simulation can deadlock when a bug in a stop task prevents it from returning, a blocked call never unblocks, or a forever loop never breaks. To prevent the simulation from hanging indefinitely, OVM provides two watchdog timeout mechanisms. One is for task phases, and the other is for stop tasks.

ovm_root contains two variables, phase_timeout and stop_timeout. Their type is the Verilog type time, and their values can be set by set_global_timeout and set_global_stop_timeout. The default value for both variables is 0, which means timeout is disabled.

When a task-based phase is executed, such as run(), and phase_timeout has been set to a value greater than zero, then a separate watchdog process is spawned that simply waits until the timeout expires. A fork/join_any construct is used to spawn these tasks, so if the run tasks finish before the timeout expires, then the timeout is ignored. On the other hand, if the timeout expires first, then it will initiate a shutdown. The code fragment in ovm_root::run_global_phase() that manages the execution of the run tasks and the shutdown is this:

```
fork : task_based_phase

  m_stop_process();

  begin
    m_do_phase_all(this,m_curr_phase);
    wait fork;
  end

  #timeout ovm_report_error("TIMOUT",
    $psprintf("Watchdog timeout of '%0t' expired.", timeout));

join_any
disable task_based_phase;
```

The fork has three processes, including m_stop_process(), which manages the stop requests; m_do_phase_all(), which causes all the run tasks to be spawned in parallel; and the timeout.

The disable statement after the join_any causes any remaining processes, whatever they happen to be, to be killed. So if the timeout expires first, then both the stop process and the run tasks will be killed. If the run tasks finish first, then the stop process and the timeout will be killed. Finally, if global_stop_request() is called, the stop tasks are called, they all complete, and then the stop process will finish first and the run tasks and timeout will be killed.

4.7   Connecting Testbenches to Hardware

Ultimately, all the class-based components must communicate with RTL hardware. SystemVerilog provides interfaces for connecting hardware objects without having to do so pin-by-pin. Hardware, in this case, means RTL components represented using Verilog modules. The language also provides virtual interfaces as a means for class-based objects to connect to RTL components. Essentially, a virtual interface is a pointer (reference) to an interface.

Figure 4-10 Interface Connecting Testbench to Hardware

To connect class-based testbench components to hardware, you must connect the hardware to an interface and then pass a virtual interface into the class-based environment. Below is an example of a pin-level interface. It is a memory interface that has an address (address), output data (wr_data), input data (rd_data), a pin that selects the data direction (rw), request and acknowledge pins (req and ack), a reset pin (rst), an error indicator (err), and of course, a clock (clk).

```
25      interface pin_if (input clk);
26         bit [15:0] address;
27         bit [7:0]  wr_data;
28         bit [7:0]  rd_data;
```

```
29        bit rst;
30        bit rw;
31        bit req;
32        bit ack;
33        bit err;
34
35        modport master_mp (
36          input   clk,
37          input   rst,
38          output address,
39          output wr_data,
40          input   rd_data,
41          output req,
42          output rw,
43          input   ack,
44          input   err );
45
46        modport slave_mp (
47          input   clk,
48          input   rst,
49          input   address,
50          input   wr_data,
51          output rd_data,
52          input   req,
53          input   rw,
54          output ack,
55          output err );
56
57        modport monitor_mp (
58          input   clk,
59          input   rst,
60          input   address,
61          input   wr_data,
62          input   rd_data,
63          input   req,
64          input   rw ,
65          input   ack,
66          input   err );
67      endinterface
file: 04_OVM_mechanics/10_vif/top.sv
```

The interface is composed of several parts. We'll look at them individually.
The first part is the header that identifies the name of the interface, in this case
pin_if. Just below that is the pin bundle that serves as the external view of
the hardware. The rest of the interface construct contains the declaration of
three *modports*. Each modport is a view of the pins in the bundle. In our
example, each modport contains all the pins in the bundle but with different
signal directions. The master modport drives transactions on the bus, so
address, wr_data, req, and rw are all outputs. The device that uses this
modport will drive those pins. The rest of the signals are inputs. Slave devices

use the slave modport, whose signal directions are set opposite the master. Monitor devices are passive and do not drive any signals. All of their signals are inputs. By choosing the appropriate modport for any device, we can easily establish the direction of all the signals and guarantee consistency across all devices connected to the bus.

In the top-level module we statically instantiate the clock generator, the interface, and the DUT.

```
150    module top;
151
152       clkgen ck(clk);
153       pin_if pif(clk);
154       dut d(pif.slave_mp);
155
156       env e;
157
158       initial begin
159          e = new("env");
160          e.set_vif(pif.master_mp);
161          run_test();
162       end
163
164    endmodule
file: 04_OVM_mechanics/10_vif/top.sv
```

The initial block dynamically instantiates the class-based testbench environment, passes the interface handle (otherwise known as a virtual interface) to the newly instantiated environment, and starts running the test. Notice that the slave modport is passed to the DUT. This is appropriate since the DUT is a memory slave. The master modport is passed into the testbench environment and ultimately to the driver. The environment stores the virtual interface and passes it to any subordinate components that may need it.

```
108    class env extends ovm_env;
109
110       local virtual pin_if vif;
111       driver d;
112
113       function new(string name, ovm_component parent = null);
114          super.new(name, parent);
115       endfunction
116
117       function void build();
118          d = new("driver", this);
119          d.set_vif(vif);
120       endfunction
121
```

```
122     task run();
123       #100;
124       global_stop_request();
125     endtask
126
127     function void set_vif(virtual pin_if _if);
128       vif = _if;
129     endfunction
130
131   endclass
file: 04_OVM_mechanics/10_vif/top.sv
```

Notice that the virtual interface, vif, is stored as a local variable. Just like any other variable, making it local prevents any unauthorized access to it. Thus, access to the interface is controlled. The set_vif() function provides the access necessary to set the value of the local virtual interface. Like the top-level environment, the driver also has a set_vif() function, which operates in precisely the same way.

```
72   class driver extends ovm_component;
73
74     local virtual pin_if vif;
75
76     function new(string name, ovm_component parent);
77       super.new(name, parent);
78     endfunction
79
80     function void set_vif(virtual pin_if _if);
81       vif = _if;
82     endfunction
83
84     task run;
85       forever begin
86         @(posedge vif.clk);
87         ovm_report_info("driver", "posedge clk");
88         //...
89       end
90     endtask
91
92   endclass
file: 04_OVM_mechanics/10_vif/top.sv
```

Hierarchically calling set_vif() functions works fine for small designs or situations where you are passing the virtual interface only one or two levels deep. In situations where you will pass the virtual interface through more levels, or more importantly, you don't know *a priori* where in the hierarchy the recipients of the virtual interface will reside, there is a more generalized way to pass virtual interfaces.

For this technique, which we call the interface object technique, create a special object to hold the interface, and pass that object to its destination using the configuration facility. The special object must be derived from ovm_object for it to be accepted by the configuration facility.

```
72    class pin_vif extends ovm_object;
73
74      virtual pin_if m_vif;
75
76      function new(virtual pin_if vif);
77        m_vif = vif;
78      endfunction
79
80    endclass
file: 04_OVM_mechanics/11_vif/top.sv
```

The class simply contains a virtual interface of the appropriate type and a constructor that sets its value, the latter being a convenience and not strictly required. In the top-level module, we create an instance of the object, assign the virtual interface, and put it into the configuration database by calling set_config_object().

```
175   module top;
176
177     clkgen ck(clk);
178     pin_if pif(clk);
179     dut d(pif.slave_mp);
180     pin_vif vif;
181
182     env e;
183
184     initial begin
185       vif = new(pif);
186       set_config_object("*", "vif", vif, 0);
187       e = new("env");
188       run_test();
189     end
190
191   endmodule
file: 04_OVM_mechanics/11_vif/top.sv
```

The set_vif() function and the local virtual interface are no longer needed in the environment. Other than the top-level module, the only component that needs to know about the interface object and the interface is the one that needs to use it. In our example, that is the driver.

```
85    class driver extends ovm_component;
```

```
86
87        local virtual pin_if vif;
88
89        function new(string name, ovm_component parent);
90          super.new(name, parent);
91        endfunction
92
93        function void build();
94
95          ovm_object dummy;
96          pin_vif v;
97
98          if(!get_config_object("vif", dummy, 0)) begin
99            ovm_report_error("get interface",
100             "no virtual interface available for driver");
101         end
102         else begin
103           if(!$cast(v, dummy)) begin
104             ovm_report_error("interface cast",
105               "supplied object is not the correct type");
106           end
107           else begin
108             ovm_report_info("get interface",
109               "interface successfully retrieved");
110             vif = v.m_vif;
111           end
112         end
113       endfunction
114
115       task run;
116         forever begin
117           @(posedge vif.clk);
118           ovm_report_info("driver", "posedge clk");
119           //...
120         end
121       endtask
122
123     endclass
file: 04_OVM_mechanics/11_vif/top.sv
```

While we are able to get rid of the local virtual interfaces and the set_vif
functions, some extra code is required in build() to retrieve the interface
object from the configuration database and make sure that it is the correct
type. We retrieve a dummy object from the configuration database using
get_config_object(). Then, if the object exists, we cast it to the type of the
interface object. If the cast succeeds, then we can reach into the interface
object to get the virtual interface and assign it to our local virtual interface.

The interface object technique of assigning virtual interfaces to components is
slightly more verbose. It requires you to create an object and put it in the

configuration database. The recipient has to retrieve the object and check to make sure that the object does indeed exist and is of the correct type. However, it is much more general and secure than hierarchical calls to set_vif(). For one thing, only the components that care about the interface must go through the extra work of retrieving the interface object. No other components have to do anything. Whereas, when using hierarchical calls to set_vif(), all components between the top level and the ones that will use the interface must store a local copy of the virtual interface and forward it downwards. Any break in the chain means the recipient will not have an interface to use.

For designs that are small, have shallow hierarchies, or only a single virtual interface to worry about, the trade-offs are not so obvious. You can successfully argue that the set_vif() technique is equivalent to or even easier than the interface object technique. However, when your design has multiple virtual interfaces and deep hierarchies, the interface object technique is clearly superior. As we will see in later chapters, using the configuration facility greatly increases the reusability of your components.

4.8   Tests and Testbenches

Through the proper use of configuration, the factory, and the phased build process, you can create a verification testbench that allows you to randomize more than just the generated stimulus. For example, if a testbench is written to allow the number of drivers on a bus to be configurable, then the same testbench can be reused across multiple *tests*, each of which might specify a different (possibly random) number of drivers. As you can see, the flexibility of OVM allows you to run each of these different tests without having to modify the testbench itself.

The OVM also provides an explicit ovm_test class as a container for tests. Typically, the top-level module will instantiate an ovm_test, which in turn configures and instantiates the testbench. Additional tests can then be written as extensions of the base test that include new configuration and factory directives, making the tests themselves relatively short, well-defined, and easy to maintain. In actuality, the ovm_test is simply another extension of ovm_component. Since tests and testbenches are simply components, they too can be created and overridden via the factory.

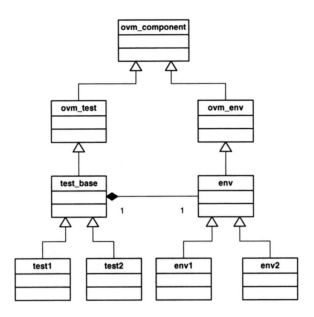

Figure 4-11 Layering Tests and Testbenches

The UML diagram above illustrates the relationship between the test and the environment. Tests and environments (env) are both components. A test contains an environment. The environment contains the top-level testbench components and their connectivity. For a particular environment, you may wish to have multiple tests. Similarly, for a particular test, you may wish to exercise it on variations of your environment. The factory lets you swap tests, environments, or both.

4.9   Reporting

OVM provides a rich set of classes and functions for generating and filtering messages. The OVM reporting facility contains three kinds of functionality:

- Displaying messages in a uniform way to various destinations
- Filtering messages
- Altering control flow as a result of a message being printed

4.9.1   Basic Messaging

`ovm_component` is a *report object*, meaning it inherits from `ovm_report_object`. `ovm_report_object`, derived from `ovm_object`, is a base class that contains all the functions you will use to issue and control messages. The four primary functions for issuing messages are:

```
function void ovm_report_info( string id,
                               string message,
                               int verbosity = OVM_MEDIUM,
                               string filename = "",
                               int line = 0);

function void ovm_report_warning( string id,
                                  string message,
                                  int verbosity = OVM_MEDIUM,
                                  string filename = "",
                                  int line = 0);

function void ovm_report_error( string id,
                                string message,
                                int verbosity = LOW,
                                string filename = "",
                                int line = 0);

function void ovm_report_fatal( string id,
                                string message,
                                int verbosity = OVM_NONE,
                                string filename = "",
                                int line = 0);
```

Each of these four functions issues a message that has several components: severity, verbosity level, identifier, message, filename, and line number.

Severity. The severity of the message can be OVM_INFO, OVM_WARNING, OVM_ERROR, or OVM_FATAL. The choice of severity changes the final text that is printed to include an indication of the severity. It also affects how the message is processed. For example, a call to ovm_report_fatal terminates the testbench. Other ways in which severity affects message processing are discussed in Section 4.9.2.

Identifier. The identifier of a message is an arbitrary string that is used to identify the string. The identifier is printed as part of the message text, and it also affects how messages are processed.

Message. The message is the body of the message text.

Verbosity. The verbosity level of a message is an arbitrary number that is relative to the current setting of the verbosity threshold. Messages whose verbosity level is at or below the threshold will be printed, and those above will be ignored. This is a way to filter messages. You can make your testbench more verbose by raising the threshold or less verbose by lowering the

threshold. The function for changing the verbosity threshold is
`set_report_verbosity_level(int verbosity)`.

Filename and line number. These are optional arguments whose role is to provide file and line number information about where the message occurred.

4.9.2 Message Actions

Associated with each message is an action that determines exactly how it is processed. The action is a bit vector with each bit representing one possible action. You can specify multiple actions by turning on one or more bits in the vector. So you don't have to remember which bit is which, OVM has an action enum that you can use to specify actions. The following table describes the possible actions:

Action	Definition
NO_ACTION	Do not execute an action.
OVM_DISPLAY	Display the message on the standard output device.
OVM_LOG	Send the message to a file.
OVM_COUNT	Increment quit_count. When quit_count reaches a predetermined threshold, terminate the testbench.
OVM_EXIT	Terminate the testbench immediately.
OVM_CALL_HOOK	Call the appropriate hook function.
OVM_STOP	Call $stop after the message has been processed.

quit_count and max_quit_count are stored in a global location. You can change max_quit_count with the following function:

```
set_max_quit_count(int q);
```

A combination of a message's severity and identifier determine the action it takes. The message handler keeps a set of tables that define actions and file destinations for messages by identifier and severity. (We'll see shortly how those tables are set up.) First, the message handler looks to see if there is an action specified for the combination of identifier and severity for the message. If there is none, then the message handler looks to see if there is an action specified just for the identifier. If it finds none, then it looks for actions by severity. The OVM message facility guarantees that there is always an action for each severity. The default actions are shown in the following table.

Severity	Default Action
OVM_INFO	OVM_DISPLAY
OVM_WARNING	OVM_DISPLAY
OVM_ERROR	OVM_DISPLAY | OVM_COUNT
OVM_FATAL	OVM_DISPLAY | OVM_EXIT

The only default actions are those determined by severity, as shown in the table above. You must set any other action by identifier or the combination of identifier and severity with functions designed for just that purpose.

4.9.3   Message Files

To send messages to a file, you must first open the file and change the appropriate message actions to OVM_LOG. A handy place to do this is in the build() method of a component, for example:

```
class component extends ovm_component;

    FILE f;

    function void build();
        f = $fopen("logfile", "w");
        set_report_default_file(f);
        set_report_severity_action(OVM_INFO, OVM_LOG);
        set_report_severity_action(OVM_WARNING, OVM_LOG);
        set_report_severity_action(OVM_ERROR, OVM_LOG);
        set_report_severity_action(OVM_FATAL, OVM_LOG | OVM_EXIT);
    endfunction
```

Later, when the testbench terminates, you can close the file:

```
    function void report();
        $fclose(f);
    endfunction
```

4.9.4   Message Handlers

Each report object has a report handler (ovm_report_handler) associated with it. The report handler is not directly accessible by the user, although it contains local state data for that report object. The report object contains no reporting data itself, only the reporting interface, that is, the functions whose work is delegated to the handler. To illustrate this concept, let's look at the hierarchical connectivity example that we discussed in Section 4.2.1. It, like all hierarchies of components, has a report handler associated with each component.

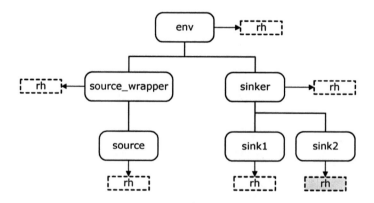

Figure 4-12 Hierarchical Design with Report Handlers

To change reporting characteristics for an individual component, you need to change only its report handler. For example, issuing this call in the component sink2:

```
set_report_id_action("fsm", OVM_LOG);
```

causes all the messages whose identifier is "fsm" to be logged to a file. Since this call was made within sink2, it only affects messages issued from sink2. Messages issued from any other component in this testbench are not affected, even if they also have the "fsm" identifier. To make a similar change on an entire sub-hierarchy, you can issue the same call on each component, or you can call the hierarchical equivalent of the set_report_id_action() method. In this case, you would call:

```
set_report_id_action_hier("fsm", LOG);
```

If you make this call in `sinker`, you will affect `sinker` and all of the components in the hierarchy beneath it. In the figure below, the shaded report handlers are those affected.

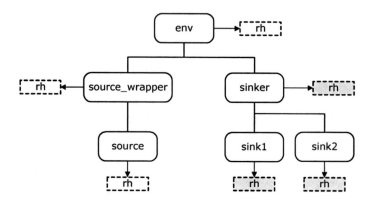

Figure 4-13 Affect of a Call to set_report_id_action_hier

The following table identifies all the methods for changing report actions and files and their hierarchical equivalents.

Local Method	Hierarchical Method
set_report_verbosity_level	set_report_verbosity_level_hier
set_report_default_file	set_report_default_file_hier
set_report_severity_action	set_report_severity_action_hier
set_report_id_action	set_report_id_action_hier
set_report_severity_id_action	set_report_severity_id_action_hier
set_report_severity_file	set_report_severity_file_hier
set_report_id_file	set_report_id_file_hier
set_report_severity_id_file	set_report_severity_id_file_hier

4.9.5 Altering the Flow of Control

Most of the time when you issue a report, the report is displayed or sent to a file, and then control resumes at the next sequential statement. There are occasions when you'll want to alter the flow of control based on a message that is issued. The most obvious case is terminating the testbench. The EXIT action terminates the testbench immediately after the message is sent to its final destination. The action COUNT increments quit_count, and the testbench terminates when quit_count reaches max_quit_count. Typically, you will use these actions to do things like prevent an errant program from looping indefinitely in an error state, or prevent cascading error messages from obfuscating the source of an error.

```
function void build();
    set_report_max_quit_count(10);
    set_report_severity_action(OVM_ERROR,
                     OVM_DISPLAY | OVM_LOG | OVM_COUNT);
endfunction
```

The example build() function above sets max_quit_count to 10 and instructs the report handler so that each time an error is issued (that is, ovm_report_error() is called) the message displays on the screen, goes to a log file, and increments quit_count. The tenth time an error is issued, the testbench terminates.

Another way to alter the flow of control when a report is issued is through report *hooks*. The report object provides this set of virtual functions. They provide a place where you can gain control when any report is issued or a report of a specific severity is issued to do additional filtering, counting, sanity checking, and so forth. The OVM report object provides five report hooks, one for each severity, and a catch-all hook that is called no matter what the severity of the report.

```
virtual function bit report_hook( string id,
                        string message,
                        int verbosity,
                        string filename,
                        int line);

virtual function bit report_message_hook( string id,
                            string message,
                            int verbosity,
                            string filename,
                            int line);

virtual function bit report_warning_hook( string id,
                            string message,
                            int verbosity,
                            string filename,
                            int line);

virtual function bit report_error_hook( string id,
                            string message,
                            int verbosity,
                            string filename,
                            int line);
```

```
virtual function bit report_fatal_hook( string id,
                                string message,
                                int verbosity,
                                string filename,
                                int line);
```

The first thing you might notice is that these functions take exactly the same argument as the ovm_report_* functions. The reason is that all of the arguments passed to ovm_report_* are passed to the hooks as well.

The other thing to notice is that each of these functions returns a value, a single bit. Processing continues only if both hooks return 1. The default hooks, the hooks in the base class that are called when you don't explicitly supply one, always return 1. If the return value is 0, then processing terminates, and it is as though the report was never issued. Through the return code of the hooks, you can do fine-grained filtering of messages. As an example of how you might use return codes, let's say that you don't want to see messages from your testbench during initialization, which takes 250 microseconds. After initialization is complete, you want to see all messages.

```
function bit report_hook(input string id,
                         input string mess,
                         input verbosity,
                         string filename,
                         int line);
    return ($time > 250000);
endfunction
```

The catch-all hook is called first, and then the severity-specific hook is called.

To enable hooks, you must turn them on by setting the action to OVM_CALL_HOOK. A convenient place to do that is in the build() function:

```
class component extends ovm_component;

    FILE f;

    function void build();
        f = $fopen("logfile", "w");
        set_report_default_file(f);
        set_report_severity_action(OVM_INFO,
                                OVM_LOG | OVM_CALL_HOOK);
        set_report_severity_action(OVM_WARNING,
                                OVM_LOG | OVM_CALL_HOOK);
        set_report_severity_action(OVM_ERROR,
                                OVM_LOG | OVM_CALL_HOOK);
        set_report_severity_action(OVM_FATAL,
                                OVM_LOG | EXIT | OVM_CALL_HOOK);
```

```
endfunction
```

Hooks are run in the component in which they are implemented. Just as each component has its own set of methods, they also have their own hooks. If you want to run the same hook in different components, you'll have to implement it in each component. A straightforward way to do this is to create your own component base class that inherits from `ovm_component` and that has your hook implementations.

4.10   Summary

An understanding of the concepts discussed in this chapter enables you to construct the essential elements of a testbench using the OVM. You can create arbitrary hierarchies of class-based components, connect them, configure them, run them, and shut them down. Subsequent chapters build upon these concepts with additional ones, with explanations along the way for creating highly reusable testbench structures.

5

Testbench Fundamentals

To answer does-it-work questions, we need to stimulate the design with known stimulus and determine if the design responds as intended. That is, we need to *control* and *observe* the DUT.

5.1 Drivers and Monitors

Two of the most fundamental objects in testbenches are *drivers* and *monitors*. A driver converts a stream of transactions into activity on a pin-level interface. A monitor does the opposite; it converts activity on a pin-level interface into a stream of transactions. Drivers are used to control the DUT by applying stimulus, and monitors are used to observe the responses.

To understand how to build and use drivers and monitors, we will start with a transaction-level example that illustrates a stimulus generator (memory master) connected with a memory slave. The memory slave is a stand-in for a driver. Whereas a true driver has a pin-level connection, the memory slave does not. What the memory slave and driver have in common is that their transaction interfaces and their internal architecture are the same. After building an understanding of the transaction-level example, we will expand it to use pin-level communication.

The memory master generates a stream of request transactions and sends them through the transport channel to the memory slave. The slave processes each request and generates responses, which it sends back to the master through the transport channel.

M. Glasser, *Open Verification Methodology Cookbook*, DOI: 10.1007/978-1-4419-0968-8_5,
© Mentor Graphics Corporation, 2009

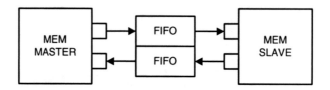

Figure 5-1 Transaction-Level Memory Master and Slave

First, let's look at the plumbing of this example, that is, the connectivity and data flow through the components. The transport channel is composed of two opposing FIFOs, one for requests and one for responses, and a transport interface. The memory master connects to the transport interface on the channel. As you might recall from Section 3.4.3, the transport interface allows the memory master to guarantee that requests and responses are synchronized. The slave interface, as the name suggests, is for devices that must respond to requests; whereas, the transport interface is for devices that generate requests.

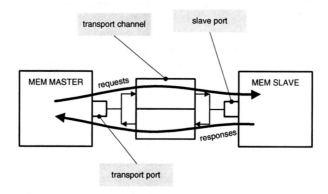

Figure 5-2 Request and Response Flow between Master and Slave

The memory master uses its port to repeatedly call transport(), which causes a request to be posted in the request FIFO of the transport channel. transport() also blocks until a response is available. The slave calls get(), which retrieves a request, then processes it, generates a response, and posts the response back to the response FIFO using put(). Finally, transport(), which has been waiting for a response, can now return the response to the memory master.

The main loop of the master contains a self-checking test. It generates a number of memory writes and saves those writes into a reference queue. Then it reads back from all the memory locations just written and compares each read value with the entry in the queue.

```
62          for(int i = 0; i < bursts; i++) begin
63            req = new();
64
65            addr = $random & addr_mask;
66            size = ($random & `h1f) +1; // size > 0 && size <= 32
67
68            // write loop
69            for(j = 0; j < size; j++) begin
70              req.set_addr(addr);
71              data = $random & data_mask;
72              refq.push_back(data);
73              req.set_wdata(data);
74              req.set_write();
75              req.set_slave_id(0);
76              transport_port.transport(req,rsp);
77              // ignore response
78              addr++;
79              #0;
80            end
81
82            // read loop
83            addr -= size;
84            for(j = 0; j < size; j++) begin
85              req.set_addr(addr);
86              req.set_wdata(0);
87              req.set_read();
88              req.set_slave_id(0);
89              transport_port.transport(req,rsp);
90              data = rsp.get_rdata();
91              refd = refq.pop_front();
92              if(data != refd) begin
93                $sformat(s, "data mismatch: %x != %x",
94                          data, refd);
95                ovm_report_error("compare", s);
96              end
97              addr++;
98              #0;
99            end
100         end
```

The main loop contains two sub-loops, a write loop and a read loop. The write loop generates a random number of writes. For each write, it generates a random data value that is both stored in the reference queue (refq) and put into the request object. transport() sends the request to the request channel and blocks until a response is available. The read loop reads back the same addresses in the same order and compares each value read with the value in the reference queue. If there is a mismatch, then an error is emitted.

The main loop of the memory slave retrieves each request, decodes it, processes it, and generates a response.

```
50          forever begin
51            slave_port.get(req);
52            assert($cast(rsp, req.clone()));
53
54            addr = req.get_addr();
55            if(req.is_read()) begin
56              data = m.read(addr);
57              rsp.set_rdata(data);
58            end
59            else begin
60              data = req.get_wdata();
61              m.write(addr, data);
62            end
63
64            slave_port.put(rsp);
65            #1;
66          end
67        endtask
```

Note that the slave uses a forever loop; whereas, the master has a bounded loop. The slave has no way of knowing up front how many requests it will process. It's the master that determines how many requests will be processed.

To send a response, we create the response object by making an exact copy of the request using clone() and then replace response fields as appropriate. This course of action is a shortcut when the request and response objects have identical types, which is the case here.

The simple producer-consumer arrangement of stimulus generator and driver is a common idiom in OVM testbenches. The simplest arrangement is a feed-forward stimulus generator that sends transactions to drive packets on a bus. More complex arrangements involve things like multiple sequences running in parallel through a sequencer to a driver. In all these cases the idea is the same: one or more testbench elements generate transactions and possibly retrieve responses connected to a driver. The driver converts the transaction stream to pin-level activity.

5.2   Introducing the HFPB Protocol

Throughout this chapter and the next several chapters, we illustrate testbench construction using a simple, non-pipelined bus protocol called the HFPB protocol. HFPB is an acronym that stands for Harry Foster peripheral bus, which is named after Harry Foster, who first suggested it. Harry drew his inspiration for the protocol from ARM's AMBA APB protocol.

The table below provides a summary of the bus signals for our simple non-pipelined bus example.

Name	Description
clk	All bus transfers occur on the rising edge of clk.
rst	An active high bus reset.
sel	These signals indicate that a slave has been selected. Each slave has its own select (for example, sel[0] for slave 0). However, for our simple example, we assume a single slave.
en	Strobe for active phase of bus.
write	When high, write access. When low, read access.
addr[7:0]	Address bus.
rdata[7:0]	Read data bus driven when write is low.
wdata[7:0]	Write data bus driven when write is high.

These signals are connected between master and slave, as illustrated in the following diagram.

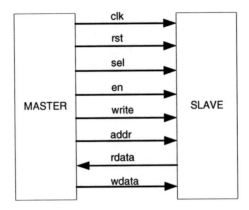

Figure 5-3 HFPB Pin Connections

The protocol operates in three states, INACTIVE, START, and ACTIVE. The relationships and transitions between the states are illustrated in Figure 5-4.

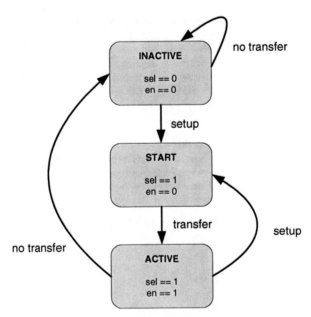

Figure 5-4 HFPB State Machine

After a reset (that is, rst==1'b1), the bus is initialized to its default INACTIVE state, which means both sel and en are de-asserted. To initiate a transfer, the bus moves into the START state, where the master asserts a slave select signal, `sel`, selecting a single slave component.

The bus only remains in the START state for one clock cycle and will then move to the ACTIVE state on the next rising edge of the clock. The ACTIVE state only lasts a single clock cycle for the data transfer. Then, the bus will move back to the START state if another transfer is required, which is indicated when the selection signal remains asserted. However, if no additional transfers are required, the bus moves back to the INACTIVE state when the master de-asserts the slave's select and bus enable signals.

The address (`addr[7:0]`), write control (`write`), and transfer enable (`en`) signals are required to remain stable during the transition from the START to ACTIVE state. However, it is not a requirement that these signals remain stable during the transition from the ACTIVE state back to the START states.

5.2.1 HFPB Write Operation

Figure 5-5 illustrates a write operation for the HFPB bus protocol involving a bus master and a single slave.

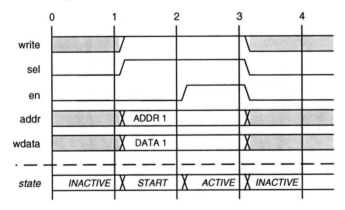

Figure 5-5 HFPB Write Transaction

At clock one, since both the slave select (`sel`) and bus enable (`en`) signals are de-asserted, our bus is in an INACTIVE state, as we previously defined in our conceptual state machine (see Figure 5-4) and illustrated in Figure 5-5. The `state` variable in Figure 5-4 is actually a conceptual state of the bus, not a physical state implemented in the design.

The first clock of the transfer is called the START cycle, which the master initiates by asserting one of the slave select lines. For our example, the master asserts `sel`, and this is detected by the rising edge of clock two. During the START cycle, the master places a valid address on the bus and in the next cycle, places valid data on the bus. This data will be written to the currently selected slave component.

The data transfer (referred to as the ACTIVE cycle) actually occurs when the master asserts the bus enable signal. In our case, it is detected on the rising edge of clock three. The address, data, and control signals all remain valid throughout the ACTIVE cycle.

When the ACTIVE cycle completes, the bus enable signal (`en`) is de-asserted by the bus master, and thus completes the current single-cycle write operation. When the master has finished transferring all data to the slave, the master de-asserts the slave select signal (for example, `sel`). Otherwise, the slave select signal remains asserted, and the bus returns to the START cycle to initiate another write operation. It is not necessary for the address data values to remain valid during the transition from the ACTIVE cycle back to the START cycle.

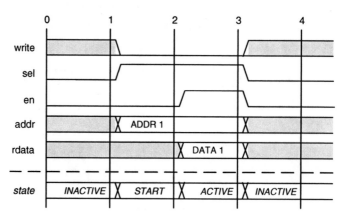

Figure 5-6 HFPB Read Transaction

5.2.2 Basic Read Operation

Figure 5-6 illustrates an HFPB read operation involving a bus master and slave zero. Just like the write operation, since both the slave select (`sel`) and bus enable (`en`) signals are de-asserted at clock one, our bus is in an INACTIVE state, as we previously defined in our conceptual state machine (see Figure 5-4). The timing of the address, write, select, and enable signals are all the same for the read operation as they were for the write operation. In the case of a read, the slave must place the data on the bus for the master to access during the ACTIVE cycle, which Figure 5-6 illustrates at clock three. Like the write operation, back-to-back read operations are permitted from a previously selected slave. However, the bus must always return to the START cycle after the completion of each ACTIVE cycle.

5.3   An RTL Memory Slave

Now, we'll expand upon our transaction-level memory master and slave example by replacing the slave with a driver and a pin-level slave. We'll also introduce a monitor.

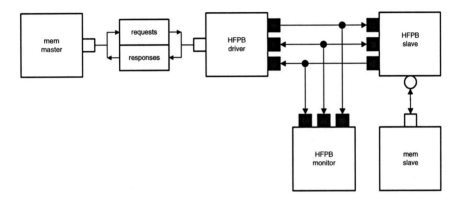

Figure 5-7 Memory Master with Driver, Monitor, and Pin-Level Slave

The first thing to notice about our expanded example is that the memory master, memory slave, and transport channel are identical to the ones from the previous example. This is an application of reuse, applying components unchanged in multiple situations.

We've inserted a pin-level driver and slave between the transaction-level master and slave. We've also added a monitor. The monitor is the complement of the driver; whereas, the role of the driver is to convert a stream of transactions into activity on the bus, the role of the monitor is to monitor the activity on the bus and convert it to a stream of transactions.

Like the transaction-level memory slave, the main loop of the driver is a forever loop. However, since the driver controls the bus, it is driven by the clock. The skeleton of the driver is based around a finite state machine coded as a case statement.   Clock control

```
      forever begin                                  Finite state machine
         @(posedge m_bus_if.master.clk)
```

```
...

    case(m_state)

      INACTIVE : begin
               ...
          end

      START : begin
               ...
          end

      ACTIVE : begin
               ...
          end

    endcase

  end // forever
```

The three-state state machine is represented using a case statement with each case containing the actions for that state. The first thing that the driver does in the INACTIVE state is get a new transaction.

```
82                if(!slave_port.try_get(m_req)) begin
83                    m_bus_if.master.sel <= 0;
84                    m_state = INACTIVE;
85                    continue;
86                end
file: 05_testbench_fundamentals/basic_hfpb/hfpb_driver.svh
```

Note that we use try_get() instead of get(). try_get() is the nonblocking variant of get(). It is a function, and therefore, it cannot consume time. If there is nothing in the FIFO to retrieve at the time it's called, try_get() returns with a status of 0. If there is something in the FIFO, it will return it along with a status code of 1. The reason we use try_get() instead of get() is because the bus is driven by the clock, and we want to ensure that remains the case. We don't want the act of getting a new request to block the bus and possibly cause other devices connected to the bus to function improperly. If try_get() does not find an item in the FIFO to retrieve then it executes an idle cycle.

In the ACTIVE state, once the transaction completes, we send a response back using try_put(). We use try_put(), which is also nonblocking, for the same reason we use try_get() — so we don't block the bus.

```
139                if(!slave_port.try_put(m_rsp))
```

```
140                  begin
141                      ovm_report_error ("MASTER",
142                                        "put response failed");
143                  end
file: 05_testbench_fundamentals/basic_hfpb/hfpb_driver.svh
```

The skeleton of the HFPB slave is the same as for the driver. This makes sense when you consider that they are both part of the same bus protocol. The actions at each state in the state machine are a bit different. Instead of driving new transactions onto the bus, as the master does, the slave responds to transactions. Once it sees a read or write transaction, it takes information from the bus and forwards it to the transaction-level memory slave. The main work done by the slave occurs in the START state.

```
78          START : begin
79
80              m_req = new();
81
82              if (m_bus_if.slave.write)
83                  m_req.set_write();
84              else
85                  m_req.set_read();
86              m_req.set_wdata(m_bus_if.slave.wdata);
87              m_req.set_addr(m_bus_if.slave.addr);
88              m_req.set_slave_id(id);
89
90              m_transport_channel.transport(m_req, m_rsp);
91
92              if(!m_bus_if.slave.write)
93                  m_bus_if.slave.rdata = m_rsp.get_rdata();
94
95          end // START
file: 05_testbench_fundamentals/basic_hfpb/hfpb_slave.svh
```

It's in the START state of the HFPB protocol that a transaction begins. The action of the HFPB (pin-level) slave is to retrieve information from the pins, determine what kind of transaction is in process, create a request object, and forward it to the transaction-level slave. The transaction-level slave does the actual processing of the request and returns a response back to the pin-level slave. The pin-level slave then puts the response on the bus once the transaction is complete. For our protocol, only reads cause the slave to change the bus pins by placing the data read from the transaction-level slave onto the bus.

5.4   Monitors and Analysis Ports

To answer the does-it-work and are-we-done questions, you must observe the DUT's behavior. You need to extract information concerning the behavior of the DUT and pass it to analysis devices dedicated to answering the relevant questions. The primary way to do this in OVM is to use *analysis ports*.

Analysis ports form the boundary between the *operational domain* and the *analysis domain* in a testbench. The analysis domain is the collection of components in the testbench responsible for analyzing the behavior observed by a monitor. *Analysis components* receive their input from analysis ports. A monitor sends transactions through an analysis port to an analysis component.

Analysis ports and analysis components together are an implementation of the *observer pattern*, a well-known, object-oriented pattern. In this pattern, the publisher provides data and the subscribers consume data. Like a magazine subscription, each subscriber must subscribe to the publisher before it can receive data. Data is transferred to the subscribers only when the publisher publishes something. Also like a magazine subscription, each subscriber receives a handle to the data from the publisher. The following diagram shows the organization of elements in an analysis port:

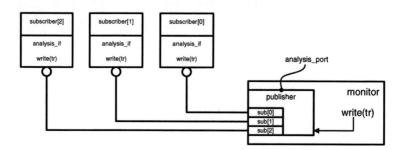

Figure 5-8 Analysis Port Organization

Before the test begins, each subscriber must register itself with the publisher. The publisher maintains a list of subscribers. At some time during its operation, the device that contains the analysis port, such as a monitor, calls write(), and passes in a transaction object. The analysis port forwards the write call to each subscriber, and passes a copy of the transaction object to the subscriber.

Analysis components (subscribers) connect to analysis ports through the *analysis interface*. The analysis interface contains the single function write().

write() is a function in SystemVerilog (not a task); therefore, it never blocks. Imagine what can happen if write() were a blocking task instead. In that case, it could interfere with the operation of the monitor. Since subscribers must receive data in the same delta cycle that the write() call is made, write() *must* be nonblocking.

Some analysis components may deliver more than one transaction through an analysis port in a single delta cycle. write() must return immediately, but the subscriber may do anything, including consume time. The consequence is that data can be lost if a subscriber is not prepared to deal with multiple transactions in one delta cycle. In this case, you can use an *analysis FIFO* to serve as a FIFO buffer between the analysis port and the analysis component. An analysis FIFO is an unbounded tlm_fifo with an analysis interface, that is, write(). Since the analysis_fifo is unbounded, write() will always be successful. The analysis component then, instead of having an analysis interface, connects to the analysis FIFO in the same way any component connects to a FIFO. It uses get() or try_get() to retrieve transactions. Of course, you can also design an analysis component that includes an analysis FIFO internally and makes the FIFO's analysis export visible as its own.

The monitor for our HFPB protocol uses the same skeleton as the driver and the slave. Through the exercise of the state machine, it is able to recognize bus transactions. As each transaction is recognized, it is sent to the analysis port using the write() call.

```
54          forever begin
55            @( posedge m_bus_if.monitor.clk );
56
57            if (m_bus_if.monitor.rst) continue;
58
59            state = state_t'({ (m_bus_if.monitor.sel != 0),
60                               m_bus_if.monitor.en });
61            case( state )
62
63              INACTIVE : begin
64              end
65
66              START : begin
67                id = 0;
68                for(id = 0; id < 8; id++)
69                  if(m_bus_if.monitor.sel[id])
70                    break;
71                if(id >= 8)
72                  id = 7;
73
74                m_trans = new();
75
76                m_trans.set_addr(m_bus_if.monitor.addr);
```

```
77                    m_trans.set_wdata(m_bus_if.monitor.wdata);
78                    m_trans.set_slave_id(id);
79
80                    if (!m_bus_if.monitor.write)
81                       continue;
82
83                    m_trans.set_write();
84                    analysis_port.write(m_trans);
85                    ovm_report_info("MONITOR", m_trans.do_sprint());
86
87                end
88
89                ACTIVE : begin
90                    if (m_bus_if.monitor.write)
91                       continue;
92
93                    m_trans.set_read();
94                    m_trans.set_rdata(m_bus_if.monitor.rdata);
95                    analysis_port.write(m_trans);
96                    ovm_report_info("MONITOR", m_trans.do_sprint());
97                end
98
99            endcase
100
101        end
file: 05_testbench_fundamentals/basic_hfpb/hfpb_monitor.svh
```

Any subscriber connected to the analysis port will receive the transaction and use it for its purposes.

5.5   Summary

We've reviewed the fundamental components of testbenches, drivers and monitors. Drivers and monitors are complementary—drivers convert transaction streams to pin wiggles, and monitors convert pin wiggles into transaction streams. Stimulus generators sending transactions to a driver and monitors observing bus activity and converting it to transactions are core idioms in testbench construction. In the next chapter, we'll look at how to apply these idioms to build complete testbenches.

6

Reuse

Building a testbench—designing, coding, debugging, and testing drivers, monitors, and other testbench components—can be quite time-consuming. An obvious place to improve verification productivity is to reuse components. That sounds simple enough, but to make a component truly reusable, some thought must be put into its architecture and construction. The types of things to think about to make a component reusable include how you expect to reuse the component and what degree of freedom the component must support.

6.1 Types of Reuse (or Reuse of Types)

The essential means to make a component reusable is to encapsulate all the data and functionality behind a well-defined interface. The interface dictates how you can modify, operate, and interrogate (extract data from) the component. All access is prohibited except that specifically allowed by the interface. We'll consider four techniques for building reusable testbench elements: function calls, parameterized classes, inheritance, and configuration. Each of these techniques represents a different way of modifying structure or behavior using an interface. In each of these techniques, information is supplied externally to change the structure or behavior of the element. The first three ways to make an element reusable are a recap from Chapter 2, where we discussed object-oriented programming.

- Function call. An algorithm or other unit of functionality is encapsulated into a function call. Whenever you need that functionality, you can simply invoke the function rather than cut-and-paste the code or rewrite it completely in place. Functions can take parameters whose values alter the behavior of the function.

M. Glasser, *Open Verification Methodology Cookbook*, DOI: 10.1007/978-1-4419-0968-8_6,
© Mentor Graphics Corporation, 2009

- Inheritance. Encapsulating data and functionality of arbitrary complexity into a single object hides that complexity so that the object can be dropped into place and operated through its interfaces. Adding to or modifying the functionality through inheritance is a way to reuse the base object and take advantage of whatever magic it contains.

- Parameterized classes provide a way to build reusable classes. A class with parameters forms a template[1] which can be instantiated multiple times with different parameters to form a family of classes. Scalar values and types can be used as parameters. Each instance of a parameterized class is called a *specialization*. To identify the specialization the parameters become part of the type.

- Run-time configuration. A configurable element can alter its behavior or structure through setting flags, switches, or configuration variables.

6.2   Reusable Components

To explore how to construct reusable components, let's consider an example of a simple memory master driving a memory through a transport channel, all at the transaction level. Let's look at each of these components in detail to see how they are constructed using reuse techniques.[2]

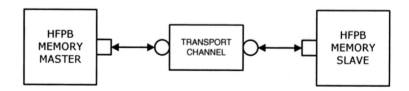

Figure 6-1 Master and Slave Connected through a Transport Channel

The class header for the memory master shows that the class is derived from another parameterized base class, `hfpb_master_base`. The base class is parameterized identically to the derived class.

---

1. What are called parameterized classes in SystemVerilog are called templates in C++. We'll use the terms template and parameterized class interchangeably.
2. Note that these examples exhibit similar topologies to those in Chapter 5. However, the components in this chapter have been designed to illustrate reuse concepts that are not discussed in Chapter 5.

```
27      class hfpb_random_mem_master
28        #(int DATA_SIZE=8, int ADDR_SIZE=16)
29          extends hfpb_master_base #(DATA_SIZE, ADDR_SIZE);
```

We've made the assumption that users of the HFPB protocol will likely build various kinds of masters to drive transactions on an HFPB bus. The hfpb_master_base allows us to put structures and functionality in the base class that will be used by all masters. So, in building our master, hfpb_random_mem_master, we reuse the functionality provided in the base class.

The HFPB master base class contains a variety of things.

```
36      class hfpb_master_base
37        #(int DATA_SIZE=8, int ADDR_SIZE=16)
38          extends ovm_component;
39
40        typedef hfpb_master_base
41          #(DATA_SIZE, ADDR_SIZE) this_type;
42        typedef ovm_component_registry
43          #(this_type) type_id;
44
45        'include "hfpb_parameters.svh"
46
47        ovm_transport_port
48          #(hfpb_tr_t, hfpb_tr_t) transport_port;
49
50        ovm_barrier objection;
51        protected hfpb_addr_map #(ADDR_SIZE) addr_map;
```

It contains a typedef of ovm_component_registry, which causes the component to be registered in the factory. It contains a transport port, an objection barrier, and an address map. These are all facilities that can be used by masters derived from this class.

The HFPB master base class is an example of using object-oriented inheritance as a reuse technique. The benefit of using inheritance is that you only have to write and test the code for facilities in the base class once. Any time you reuse the base class to build a new bus master, you are guaranteed consistency in structure of the master. For example, you will always know that masters derived from hfpb_master_base will have a transport port and its name is transport_port.

hfpb_random_mem_master, the derived master, randomizes a sequence of memory transactions. Some parameters, max_bursts and max_burst_size, guide the randomization. max_bursts is the maximum number of bursts to be issued by the master in a test, and max_burst_size is the maximum number of transactions in a single burst. Instead of hardcoding these values, we make them available to the object through the configuration facility.

```
54        max_burst_size = 16;
55        if(!get_config_int("max_burst_size", max_burst_size))
begin
56            $sformat(s, "max burst size not specified, using
default of %0d", max_burst_size);
57            ovm_report_warning("build", s);
58        end
59        $sformat(s, "max burst size: %0d", max_burst_size);
60        ovm_report_info("build", s);
61
62        max_bursts = 100;
63        if(!get_config_int("max_bursts", max_bursts)) begin
64            $sformat(s, "max bursts not specified, using default
of %0d", max_bursts);
65            ovm_report_warning("build", s);
66        end
67        $sformat(s, "max bursts: %0d", max_bursts);
68        ovm_report_info("build", s);
```

For each of the two parameters, we first establish a default value of 16 for max_burst_size and 100 for max_bursts. Then, for each one, we call get_config_int to see if a value has been specified externally. If not, we issue a warning to alert the user that no value has been specified and use the default.

It's important to issue the warning message if get_config_* returns a 0. Without the warning, the component silently uses the default value *even if that was not the intention*. It's possible, for example, that the test writer neglected to supply a value in the configuration database for max_burst_size and max_bursts. Or, in the call to set_config_int, it's possible that the test writer mistyped one of the names. In that case, not realizing the mistake, the test writer would believe that the value was being set as specified. In this case, because of the unnoticed misspelling, the get_config_int call finds no value to retrieve, and instead, it uses the default. Furthermore, the test behavior is different than intended, with no warning that anything might be amiss. With the warning, the user can later look at the test run to determine if the test behaved as intended.

This is an example of applying the config facility to make a component reusable. Instead of building separate versions of the component, each with different characteristics, we identify the characteristics that might change and provide a means for them to be modified without having to alter the component itself.

Verification components—drivers, monitors, and so forth—are typically protocol-specific, meaning they know about one and only one particular protocol. As we saw in the previous chapter, it's straightforward to build protocol-specific components. Protocols, however, often come in variations. There might be the 16-bit and 32-bit versions, or the number of masters and slaves might change, or some other characteristic of the protocol might be configurable. The HFPB protocol, for example, can have a data bus of arbitrary size, an address bus of arbitrary size, and the address and data buses do not have to be the same size. Rather than build separate components for each configuration that we might be interested in, we build a parameterized component where the data bus size and address bus size can be modified using class parameters. The code inside the component is written to be independent of the parameter. That is, it makes no (or limited) assumptions about what values the parameter takes so that anywhere the value is needed, instead of supplying a constant, you supply the parameter. The hfpb_driver is one such parameterized component.

```
23      class hfpb_driver #(int DATA_SIZE=8, int ADDR_SIZE=16)
24        extends
25          ovm_driver #(hfpb_seq_item #(DATA_SIZE, ADDR_SIZE),
26                       hfpb_seq_item #(DATA_SIZE, ADDR_SIZE));
```

The driver has two parameters, DATA_SIZE and ADDR_SIZE. Each has a default value, so if you do not supply one, the default is used. Any time you declare an object of a parameterized type, you create a specialization, a copy of the code with the parameter value substituted for its name. You don't ever see the specialized code; the compiler takes care of managing it without your intervention. The driver is derived from ovm_driver, a base class that is discussed in Chapter 8. ovm_driver is parameterized by the sequence item types passed between it and the sequencer. In the case of the HFPB protocol, the sequence items are also parameterized with DATA_SIZE and ADDR_SIZE.

The driver code is written so that any place there is a need for either the data bus size or the address bus size, DATA_SIZE and ADDR_SIZE parameters are used as constants. We can use a parameterized component in many different situations, and by doing so, we maintain independence of any specific value of either DATA_SIZE or ADDR_SIZE. Then, by altering the parameters, we can affect the component's structure. Thus, we can reuse our parameterized component in systems with different address and data bus widths.

By building all the HFPB components so that they are parameterized in the same manner, we can construct an entire testbench that is independent of the address and data bus widths. Starting at the top, we supply the ADDR_SIZE and DATA_SIZE parameters in the top-most module. This is the only place it is necessary to specify values for DATA_SIZE and ADDR_SIZE. Everywhere else, the values are received through class parameters.

```
83    module top;
84
85       parameter int DATA_SIZE = 8;
86       parameter int ADDR_SIZE = 9;
87
88       env #(DATA_SIZE, ADDR_SIZE) e;
89
90       initial begin
91         e = new("env");
92         run_test();
93       end
94
95    endmodule
file: 06_reuse/01_TL/top.sv
```

Those parameters are passed to env, the topmost testbench component, by creating a specialization of the parameterized class. env, of course, is a parameterized component whose parameters are also DATA_SIZE and ADDR_SIZE. Further, any components instantiated in env that depend on the address or data bus sizes are similarly parameterized.

```
42    class env #(int DATA_SIZE=8, int ADDR_SIZE=16)
43       extends ovm_component;
44
45       hfpb_mem #(DATA_SIZE, ADDR_SIZE) mem;
46       hfpb_random_mem_master #(DATA_SIZE, ADDR_SIZE) mm;
47       hfpb_addr_map #(ADDR_SIZE) addr_map;
48
49       tlm_transport_channel
50         #(hfpb_transaction #(DATA_SIZE, ADDR_SIZE),
51           hfpb_transaction #(DATA_SIZE, ADDR_SIZE))
52             transport_channel;
file: 06_reuse/01_TL/top.sv
```

6.3   Agents

It's quite common to find that when you build testbenches you will see a lot of repeated instantiations and connections. It is typical to connect drivers and monitors in much the same way. Also, you will often find that when you repeat certain conglomerations of components, you will want them to be consistently configured. Individually instantiating and configuring components can introduce error-prone tedium. *Agents* address this problem.

Agents are all about reuse. They reuse monitors, drivers, and other components that are part of a particular protocol, and they themselves form reusable components by creating an interface around the subordinate components. An agent contains all the elements of a protocol encapsulated in a single package. You can apply the protocol easily in a testbench by instantiating this package, rather than separately instantiating a driver, monitor, and other protocol-specific components.

The agent is the wrapper around all the components that implement a protocol. It serves as the interface to all the protocol components. The interface takes several forms: class parameters, ports and exports, a virtual interface, and a configuration interface. The class parameters are passed on to the subordinate protocol components. The ports and exports provide the ingress and egress points for transactions; the virtual interface provides the pin-level connection point; and the configuration interface enables you to turn on or off various components to customize the agent's behavior for specific applications.

Figure 6-2 shows a simple agent with just a driver and a monitor. The external interfaces include an export for a transaction ingress, an analysis port (which makes available all bus transactions), and a virtual interface for pin-level connections to the bus.

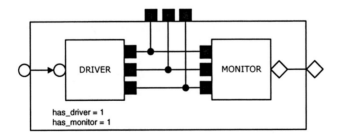

Figure 6-2 Simple Agent with a Driver and a Monitor

In some cases, you may not need both a monitor and a driver. Say you only need the driver. You can still use the agent and just turn off the monitor. The agent uses the configuration system to determine which internal components are enabled or disabled. The has_driver and has_monitor flags (in this case) are used to select which components are enabled.

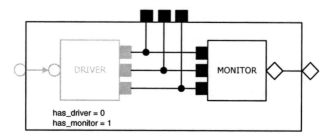

Figure 6-3 Simple Agent with Driver Disabled

The simple agent with the driver turned off functions simply as a monitor. Conversely, you can turn off the monitor and operate the device as a driver. Of course, it doesn't make sense to turn off both the driver and monitor because then the agent would not do anything.

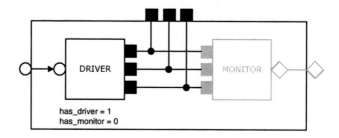

Figure 6-4 Simple Agent with the Monitor Turned Off

If you only want to use either a driver or a monitor, why bother using the agent? Why not just simply instantiate a driver or a monitor as required? The reason is simple: reuse. If we instantiate only a driver in the environment and later decide we need a monitor, then we have to get the text editor out and change the environment code to instantiate the monitor and connect it. Since we used an agent, we can just turn on the has_monitor switch and not have to modify the environment code at all.

6.4   Reusable HFPB Protocol

The HFPB agent is a highly parameterized and configurable device. It contains all of the protocol-specific components needed for the HFPB protocol in one instantiable component. This includes masters, drivers (we'll explain the difference shortly), sequencer, slaves, and a coverage collector. The connections between these components are specified in the agent. Like the simple agent discussed above, this agent has class parameters, a collection of transaction-level ports and exports, a virtual interface for pin-level connections, and a configuration interface.

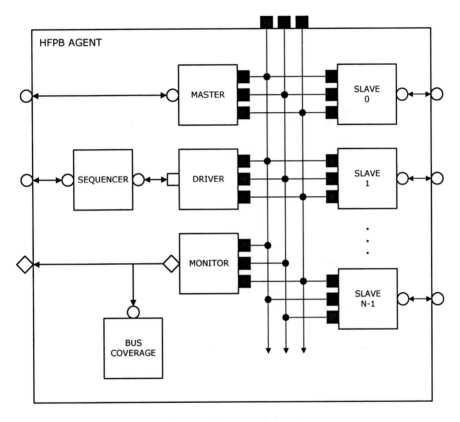

Figure 6-5 HFPB Agent

The parameters in the agent's class header, DATA_SIZE and ADDR_SIZE, are used in creating the internal structure of the agent.

```
138    class hfpb_agent #(int DATA_SIZE=8, ADDR_SIZE=16)
139       extends ovm_agent;
```

The external connections include a virtual interface for pin-level connections, a transport export for traditional TLM use, a sequence pull port for connecting to a sequencer, an array of slave exports (one for each slave), and an analysis port for transmitting transactions recognized on the pin-level bus.

```
142        virtual hfpb_if #(DATA_SIZE, ADDR_SIZE) m_bus_if;
143
144        ovm_transport_export
145          #(hfpb_transaction #(DATA_SIZE, ADDR_SIZE),
146            hfpb_transaction #(DATA_SIZE, ADDR_SIZE))
147              transport_export;
148
149        ovm_seq_item_pull_port
150          #(hfpb_seq_item#(DATA_SIZE, ADDR_SIZE),
151            hfpb_seq_item#(DATA_SIZE, ADDR_SIZE))
152              seq_item_port;
153
154        ovm_slave_export
155          #(hfpb_transaction #(DATA_SIZE, ADDR_SIZE),
156            hfpb_transaction #(DATA_SIZE, ADDR_SIZE))
157              slave_export [];
158
159        ovm_analysis_port
160          #(hfpb_transaction #(DATA_SIZE, ADDR_SIZE))
161              analysis_port;
```

The agent contains a collection of protocol components. Notice that they are all declared as local, except for the sequencer. All access to these components is through the interfaces just listed and not directly to the components. This data hiding helps us ensure that the agent remains reusable by not allowing users to form improper dependencies on the internal objects of the agent.

```
164        local hfpb_master #(DATA_SIZE, ADDR_SIZE) master;
165        local hfpb_driver #(DATA_SIZE, ADDR_SIZE) driver;
166        local hfpb_slave #(DATA_SIZE, ADDR_SIZE) slave [];
167        local hfpb_monitor #(DATA_SIZE, ADDR_SIZE) monitor;
168        local hfpb_coverage #(DATA_SIZE, ADDR_SIZE) cov;
169        local hfpb_talker #(DATA_SIZE, ADDR_SIZE) talker;
170        hfpb_sequencer #(DATA_SIZE, ADDR_SIZE) sequencer;
```

The reason the sequencer is not declared as local is because it is necessary to access the sequencer in order to operate it. In our HFPB agent we have both a master and a driver. The difference between a master and a driver is that the master has a transport export and the driver has a seq_item_pull_port for sequences, which is connected to a sequencer. (We'll discuss sequences and sequencers at length in Chapter 8). Also, the objects accepted on the input of the master must be derived from ovm_transaction and the objections accepted by the driver must be derived from ovm_sequence_item. These are two different means for moving request and response transactions into and out of the agent. The master is used for traditional structural transaction-level models, and the driver is for processing transactions (in the form of sequence items) generated by sequences. The HFBP protocol allows for exactly one master and one or more slaves, so the master and driver are mutually exclusive. It doesn't make sense in this protocol to have more than one device driving the bus.

An important thing to observe is that the interfaces and internal components have class parameters that are identical to the agent's, that is, DATA_SIZE and ADDR_SIZE. By creating a family of components for the HFPB protocol that are parameterized identically, we can pass parameters and avoid specifying the values of those parameters more than once.

Not all of the internal components and interfaces are instantiated each time the agent is used. Only those that are needed for the specified configuration are created. Rather than create separate agents for each possible configuration, which would be very clumsy, we use the configuration facility in OVM to change the structure of the agent. Our HFPB agent has a number of configuration parameters that control its structure.

```
175        local bit has_monitor;
176        local bit has_coverage;
177        local bit has_talker;
178        local bit has_master;
179        local bit has_driver;
180        local bit has_sequencer;
181        local int unsigned slaves;
182        local hfpb_vif #(DATA_SIZE, ADDR_SIZE) vif;
```

In build() and connect() the agent uses the values of these configuration parameters to control which sub-components are instantiated and how they are connected. The value for each configuration parameter is obtained through calls to get_config_int (or, in the case of vif, get_config_obj). In the agent's build() function is a series of calls to get_config_* to obtain all the configuration information needed by this component.

```
199        if(!get_config_int("has_monitor", has_monitor)) begin
200          has_monitor = 0;
201          monitor = null;
202        end
203
204        if(!get_config_int("has_coverage",
205                              has_coverage)) begin
206          has_coverage = 0;
207          cov = null;
208        end
```

. . .

As the `build()` function obtains all the configuration parameters, it also does any necessary error checking and consistency checking. Each `get_config_*` call is encased in a conditional statement that checks the status of the call. If the call fails, meaning there is no item with the specified name for the current scope in the configuration database, then we make sure the configuration parameter is set to a legal value. Other variable settings are also made, as appropriate. For example, if `has_monitor` is not supplied, then we make sure it is set to 0 and the monitor handle is null. If `has_coverage` is set to 1 we also set `has_monitor` to 1 because it would not make any sense to have a coverage collector and no monitor. And so on.

Further down in build, we use the configuration parameters to construct the internals of the agent. As an example, if `has_master` is set, then we instantiate the master component and the transport export that it will use to connect to external components.

```
284        if (has_master) begin
285            master = new("master", this);
286            transport_export = new("transport_export", this);
287        end
```

Later, in the connect phase, we will again use the `has_master` configuration parameter, this time to determine whether to connect to the transport export.

```
340        if (has_master) begin
341            transport_export.connect(master.transport_export);
342        end
```

This check is necessary because if `has_master` is 0, then we know that neither the master component nor the transport export were instantiated. Checking the value of `has_master` again ensures that we don't attempt to connect components that were not instantiated.

The virtual interface `vif` is also obtained through the configuration facility using the interface object technique described in Section 4.7. Strictly speaking, the interface object is not a configuration parameter. Virtual interfaces are part of the connectivity of the design. If you leave out the virtual interface, it's likely the agent won't work correctly in the design.

We can use our highly configurable agent in a variety of ways. The table below summarizes some of the interesting configurations

Mode	Config Settings
Monitor	has_monitor = 1 has_coverage = don't care has_talker = don't care has_master = 0 has_driver = 0 has_sequencer = 0 slaves = 0
Master	has_monitor = don't care has_coverage = don't care has_talker = don't care has_master = 1 has_driver = 0 has_sequencer = 0 slaves = 0
Driver	has_monitor = don't care has_coverage = don't care has_talker = don't care has_master = 0 has_driver = 1 has_sequencer = 0 slaves = 0
Sequencer	has_monitor = don't care has_coverage = don't care has_talker = don't care has_master = 0 has_driver = 1 has_sequencer = 1 slaves = 0

In monitor mode, all the components are turned off except the monitor. The agent then functions strictly as a monitor. You can similarly configure master mode by turning off everything except the master and then operate the agent solely as a master. You can turn on or off any component or subset of components in the agent.

By encapsulating all of the protocol-specific components in a single component called an agent, and by employing the OVM configuration facility, you can create a single component that can be used for a variety of applications. One reason for all this configurability is to enable reuse of block-level testbench components, unchanged, in a system-level testbench. We will discuss this in detail in Chapter 9.

6.5   Agent Example

As an example of using an agent, we'll transform the design shown in Figure 6-1 to use an agent instead of a transport channel. We'll also add a second memory.

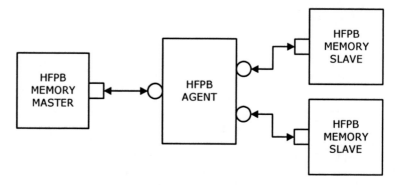

Figure 6-6 An Example Using an Agent

In this example, the agent serves as the bus model. It contains the master and slaves and all the connections between them. In addition, it contains a monitor. For this example, we configure the agent in this way:

```
80          set_config_int("hfpb_agent", "has_monitor", 0);
81          set_config_int("hfpb_agent", "has_master", 1);
82          set_config_int("hfpb_agent", "slaves", 2);
83          set_config_int("hfpb_agent", "has_talker", 0);
84          set_config_object("*",         "addr_map", addr_map, 0);
file: 06_reuse/02_RTL/top.sv
```

The monitor, master, and talker are turned on, and the bus is configured to have two slaves. Additionally, an address map is supplied, which identifies which part of the address space each memory slave occupies. A talker is a subscriber device connected to the analysis port. It simply prints out the transactions recognized by the monitor. The talker is turned on when you want to see a printed report of all the transactions that go through the agent. Given this particular configuration, we are effectively building a topology shown in Figure 6-7.

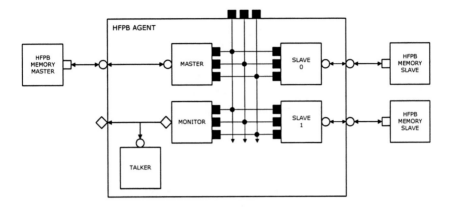

Figure 6-7 Complete Topology of Agent Example

The top-level module for this design is similar to the one described in the example in Section 6.2. Like that example, the values for the DATA_SIZE and ADDR_SIZE parameters are set here and passed into the class-based environment, which in turn, passes them on to the agent.

```
120    module top;
121
122       parameter int DATA_SIZE = 8;
123       parameter int ADDR_SIZE = 9;
124
125       env #(DATA_SIZE, ADDR_SIZE) e;
126       hfpb_vif #(DATA_SIZE, ADDR_SIZE) hfpb_vif_obj;
127
128       clk_rst cr();
129       clock_reset ck (cr);
130       hfpb_if #(DATA_SIZE, ADDR_SIZE) bus_if (cr.clk, cr.rst);
131
132       initial begin
133
134          e = new("env");
135          hfpb_vif_obj = new(bus_if);
136          set_config_object("*", "hfpb_vif", hfpb_vif_obj, 0);
137
138          fork
139             ck.run();
140          join_none
141
142          run_test();
143       end
144
145    endmodule
file: 06_reuse/02_RTL/top.sv
```

In addition, we use the interface object technique described in Section 4.7 to pass the virtual interface into the class-based environment.

Through the application of parameterized classes and the OVM configuration facility we are able to create a single component, the HFPB agent, that enables us to realize a variety of topologies and configurations of HFPB protocol components.

6.6   Summary

One of the primary keys to improving the productivity and reliability of your verification flow is reuse. Reusing components saves time by not having to write new code. Reused components are more robust by virtue of the fact that they have been used in multiple applications. Reusable components don't fall out for free; you have to put some thought into how you will reuse a particular component. Fortunately, SystemVerilog and OVM provide some facilities that encourage the development of highly reusable testbench elements.

## 7

Complete Testbenches

To answer does-it-work and are-we-done questions, we need more than just drivers and monitors. We need to collect coverage information to answer are-we-done questions; we need a reference model and a mechanism to compare the function of the reference model to answer does-it-work questions; and we need a control mechanism to shut down the testbench at the appropriate time. Finally, we need some adapters and connectors to put the whole thing together.

7.1   Floating Point Unit

This and subsequent chapters illustrate testbench construction techniques using a floating point unit (FPU) design. This design accepts a pair of floating point operands and an operator, and computes the result. This section presents an example that uses the transaction-level FPU to illustrate the construction of an OVM coverage collector. The figure below shows the organization of the example.

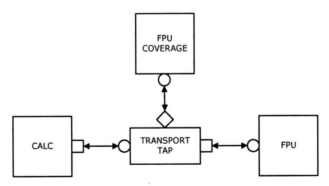

Figure 7-1 Simple FPU Testbench with Coverage

M. Glasser, *Open Verification Methodology Cookbook*, DOI: 10.1007/978-1-4419-0968-8_7,
© Mentor Graphics Corporation, 2009

The calculation generator, CALC, sends randomized arithmetic calculations to the FPU in the form of FPU request transactions. The FPU performs the calculation and returns a response transaction that contains the result of the calculation. A tap is connected between the stimulus generator and the DUT. The tap's role is to form requests and responses into pairs and send the pairs to analysis components via an analysis port.

The transaction-level FPU is a simple device. The kernel of the model is the classic transaction-level modeling idiom for slaves: getting a request, processing it, and returning a response.

```
58          forever begin
59            get_port.get(req);
60            rsp = compute(req);
61            put_port.put(rsp);
62          end
63        endtask
```

The communication interface to the FPU is handled via an embedded tlm_transport_channel (see "Transport" on page 58). To simplify coding of the run() task, we declare a get_port and a put_port to connect to the internal slave side of the transport channel. The get_port is for retrieving requests, and the put_port is for returning responses.

```
23    class fpu_tlm extends ovm_component;
24
25      ovm_transport_export #(fpu_request,
26                             fpu_response) transport_export;
27
28      local tlm_transport_channel #(fpu_request,
29                                    fpu_response) mstr_chan;
30      local ovm_blocking_get_port #(fpu_request)  get_port;
31      local ovm_blocking_put_port #(fpu_response) put_port;

44      function void connect();
45        transport_export.connect(mstr_chan.transport_export);
46
get_port.connect(mstr_chan.blocking_get_request_export);
47
put_port.connect(mstr_chan.blocking_put_response_export);
48      endfunction
```

The heart of the compute function is a case statement that performs the requested arithmetic operation.

```
75          case(op)
```

```
76              OP_ADD: result = req.a + req.b;
77              OP_SUB: result = req.a - req.b;
78              OP_MUL: result = req.a * req.b;
79              OP_DIV:
80                 if (req.b <= 1.0e-38 && req.b >= -1.0e-38)
81                    result = _nanf(); // div by zero
82                 else
83                    result = req.a / req.b;
84              OP_SQR:
85                 if (req.a < 0.0)
86                    result = _nanf();
87                 else
88                    result = req.a ** 0.5; // square root
89           endcase
```

The case statement switches on the requested operation, and each of the case branches performs a specific computation. The divide case first checks to see if the divisor is zero, since division by zero is undefined. In the RTL version of the FPU, a divide-by-zero exception will be raised when a division operation is selected and the divisor is zero. The square root branch checks to see if its operand is less than zero, since the square root of a negative number is also undefined. In both cases, the result is set to NaN, meaning not a number, which is the IEEE floating point standard value for undefined values.

The key design consideration of the transport tap is that it doesn't consume any time. That is, there must be no delta cycle delay between the transport call in the generator and the transport call in the slave. To meet this requirement, we implement the blocking transport interface directly using an ovm_blocking_transport_imp. Our implementation of transport() forwards the request downstream and the response back upstream. It also forms the request and response into a pair and sends it out an analysis port.

```
89           task transport(input fpu_request req,
90                          output fpu_response rsp);
91
92              fpu_pair pair;
93
94              transport_port.transport(req, rsp);
95              pair = new(req, rsp);
96              pair_ap.write(pair);
97
98           endtask
file: 07_complete_testbenches/01_tlm_reference/top.sv
```

By supplying an implementation instead of connecting to a channel, we avoid any delays associated with moving data through the channel.

7.2   Coverage Collectors

A key component in answering the are-we-done question is the *coverage collector*. Its role, as the name suggests, is to collect functional coverage information as the simulation proceeds. Coverage is a quantitative measure of how much of the design a test has exercised. Coverage collectors obtain information about what has been exercised and use it to calculate the answer to are-we-done questions.

Coverage collectors are constructed in OVM as extensions of the ovm_subscriber abstract base class. As you can see below, the subscriber has an implementation of the analysis interface that contains a single nonblocking function, write().

```
virtual class ovm_subscriber #(type T = int)
  extends ovm_component;

  typedef ovm_subscriber #(T) this_type;

  ovm_analysis_imp #(T, this_type) analysis_export;

  function new(string name, ovm_component parent);
    super.new(name, parent);
    analysis_export = new( "analysis_imp", this );
  endfunction

  pure virtual function void write( input T t );

endclass
```

Since it is an abstract class, you must extend ovm_subscriber and define your own implementation of the write() function to record the coverage. The ovm_analysis_imp binds the connector to the actual interface implementation. We call it analysis_export because, externally, it looks exactly like an export, in that it provides an implementation to the calling port. When you extend ovm_subscriber, you simply connect the analysis_export to the desired analysis_port, and you're in business.

The implementation of the write() function collects data from the object passed in as its argument and processes it. The processing can be of any sort, as long as it maintains the nonblocking semantic. SystemVerilog provides the *covergroup* construct to aggregate and process the actual coverage data. Usually, the write() method will copy relevant fields of its input transaction into a class variable that is then sampled by the covergroup. In effect, the role of the subscriber is to provide a means to connect the covergroup to other OVM components that feed the subscriber data to analyze.

In our FPU example, the coverage collector that is connected to the tap serves a dual role. It collects coverage, of course, *and* it shuts down the simulation when a coverage threshold is reached.

The work of the coverage collection is performed by a covergroup embedded in the subscriber. This particular covergroup, called fpu_cov, has a single coverpoint, cons_op, which counts FPU operations. For this particular covergroup, 100 percent coverage is reached when each operation is executed twice.

```
57      covergroup fpu_cov;
58         cons_op : coverpoint m_op {bins adds = (OP_ADD [* 2]);
59                                    bins subs = (OP_SUB [* 2]);
60                                    bins muls = (OP_MUL [* 2]);
61                                    bins divs = (OP_DIV [* 2]);
62                                    bins sqrs = (OP_SQR [* 2]); }
63      endgroup
```

Of course, the implementation of covergroups and the criteria for reaching full coverage is application-specific. This particular covergroup is used to illustrate the concept.

The write() function has three responsibilities in our coverage collector. It copies data from the transaction passed into a class variable so the data is visible to the covergroup, it calls sample() on the covergroup, and it tests to see if the coverage threshold has been reached. The call to sample() instructs the covergroup to look at the current values of the relevant covered variables and update its counts.

```
78      function void write(input fpu_pair t);
79
80         real coverage;
81         m_op = t.req.op;
82         m_round = t.req.round;
83
84         fpu_cov.sample();
85
86         coverage = fpu_cov.get_inst_coverage();
87         if(coverage >= coverage_threshold) begin
88            done = 1;
89         end
90
91      endfunction
```

In addition to keeping count of coverage information, the coverage collector allows the testbench to be shut down when the coverage threshold is reached.

It does this in conjunction with the top-level environment. The top-level environment calls `global_stop_request()` as the first (and only) statement in its `run()` task. This causes `stop()` to be called in all components that have `enable_stop_interrupt` set to 1, including our coverage collector. When all `stop()` tasks return, then the simulation shuts down. Our coverage collector does not return from the `stop()` task until the `done` bit is set, indicating that full coverage has been reached.

```
96      task stop(string ph_name);
97        wait (done == 1);
98        ovm_report_info("stop", "allowing stop");
99      endtask
```

Notice that we do not need any explicit communication path between the coverage collector and the test. The `stop_request` mechanism in OVM automatically handles the proper notification. The test will not complete until all components in the testbench (that have their `enable_stop_interrupt` bit set) return from their `stop()` methods, or a timeout has been reached.

So, just because the `stop()` task in this component is unblocked, it doesn't mean the simulation will immediately terminate. It only means that from the perspective of this coverage collector, it's okay for the simulation to terminate. It could be that there are other components whose `stop()` task still blocks for one reason or another. Only when all the `stop()` tasks return will shutdown begin.

7.3   FPU Agent

As we refine the transaction-level model of the FPU to RTL, it becomes necessary to ensure that the interface protocol is exercised fully and correctly. To illustrate this verification task, we use an RTL version of the FPU model written in VHDL.[1] The interface to the FPU is straightforward, which makes it a good candidate for the examples. Figure 7-2 shows the pinout for the FPU block. It has two 32-bit input buses for A and B operands and a 32-bit bus for the output result. A 2-bit input bus defines the rounding mode, and a 3-bit input bus defines the operation to be performed. Eight output pins signal exceptions, one per pin

---

1. The FPU in our examples is the FPU100 design from opencores.org. For complete details see http://www.opencores.org/projects.cgi/web/fpu100/overview.

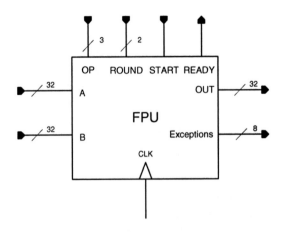

Figure 7-2 Pinout for FPU

The 32-bit A and B operands and the 32-bit result are floating point values and are represented using the IEEE 754 standard for binary representation of floating point values.

The tables below summarize the function of the FPU.

Exceptions		Round		Operation
inexact	00	even	000	add
overflow	01	zero	001	subtract
underflow	10	up	010	multiply
divide-by-zero	11	down	011	divide
infinite			100	square root
zero			101	unused
Qnan			110	unused
Snan			111	unused

The FPU is operated by the start pin. A calculation begins on the next rising clock edge when the start pin is asserted. The FPU asserts the ready pin when the calculation is complete. The device is pipelined with a depth of 1 so when the ready pin asserts, it indicates that the result of the previous calculation is available on the outputs.

To use the FPU in testbenches it's convenient to treat it like a protocol—to create a driver, monitor, and so forth, and encapsulate them in an agent. While the interface to the FPU is not a general-purpose protocol like our HFPB protocol or others such as USB, PCI, and so forth, thinking of it in those terms enables us to create reusable components for building testbenches for the FPU or devices that use the FPU.

The organization of the FPU agent is shown in Figure 7-3 below. It's organized much like the HFPB agent. It has masters and drivers, which convert transactions into pin-level activity. It also has a pin-level bus, monitor, talker, and coverage collector. One important difference is that it doesn't have slaves, so you can't really use the agent as a standalone bus model. Since the FPU interface is a protocol for accessing a specific device and not a bus or a communication protocol, there is no problem.

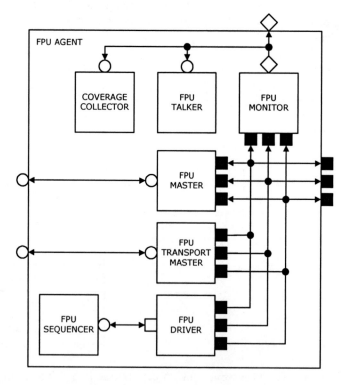

Figure 7-3 FPU Agent

As with our HFPB Agent, the FPU Agent includes a protocol-specific coverage collector. The monitor detects request and response transactions on the pin-level interface and assembles them into an `fpu_pair` transaction, which includes both the request and response transactions embedded in it. One way to customize the agent is to configure it via the factory to instantiate different coverage collectors, depending on what you are trying to accomplish in your test.

The FPU agent has three ways to drive transactions on the bus: a master, a transport master, and a sequencer-driver combination. The master and transport master use transaction objects derived from `ovm_transaction`. The sequencer-driver uses transaction objects derived from `ovm_sequence_item`. The difference is that sequence items have some extra machinery that enables them to be transported through a sequencer to a driver. Besides being derived from different base classes, the contents of the sequence items and the transactions are identical. Chapter 8 discusses sequences and sequence items more thoroughly. These three components for driving transactions to the FPU are mutually exclusive. Good coding practices dictate that the configuration interface for the FPU agent ensures that no more than one form is instantiated.

7.4   Scoreboards

The term *scoreboard* is a generic term for a wide range of component types whose function is to answer does-it-work questions. The essential characteristic of a scoreboard is that it collects data about the operation of the DUT as the simulation proceeds and compares it with a reference of some sort to determine if the DUT is functioning correctly. A scoreboard can be as simple as a trigger that recognizes when a flag is raised or a truth table, as in the simple testbench in Section 1.2.2. Or it can be as complex as a complete reference model of a complete system design.

For the FPU design, we embed a scoreboard inside a reference model. The The reference for the FPU contains the transaction-level implementation of the FPU and a scoreboard to compare the data generated by reference with the results from the RTL DUT. Figure 7-4 shows the reference, along with an example design that uses it.

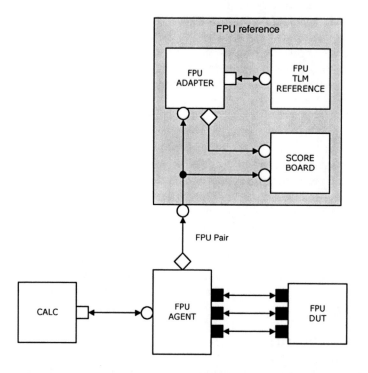

Figure 7-4 Complete FPU Testbench

In this application, the reference model serves as an analysis component. That is, the agent sends request-response transaction pairs (fpu_pair) out its analysis_port to be used for additional analysis beyond the protocol-specific coverage calculated in the agent. Since an analysis_port is unidirectional but the TLM FPU model is bidirectional, we create an adapter component that interfaces between the two.

The FPU adapter includes an analysis_fifo (see Section 5.4) to hold the fpu_pair transactions coming from the agent. The run task of the adapter gets the transaction pair from the FIFO, extracts the request transaction, and sends it to the TLM reference model, which returns a response transaction. The adapter writes this reference response, along with the request, as a new pair to its analysis port.

```
46        task run;
47
48            fpu_pair in_pair;
49            fpu_pair out_pair;
50            fpu_response rsp;
```

```
51
52          forever begin
53            pair_fifo.get(in_pair);
54            transport_port.transport(in_pair.req, rsp);
55            out_pair = new(in_pair.req, rsp);
56            pair_ap.write(out_pair);
57          end
58        endtask
```

The scoreboard has two analysis exports, one connected to the agent and the other connected to the adapter. Its run task continually compares the reference response to the original response from the agent and reports any errors.

7.5   Different Tests

Once the basic testbench topology has been defined, OVM lets you define a test class that instantiates the testbench and optionally modifies it via the factory or configuration mechanisms. Since tests themselves are classes, it is easy to create additional tests as straightforward extensions of a base test. The test to be run can simply be specified on the command line via the OVM_TESTNAME plusarg or as a string argument to the run_test() task.

Since a test always instantiates a testbench, the instantiation is typically done in the base test.

```
110    module top;
111
112      parameter int DATA_SIZE = 8;
113      parameter int ADDR_SIZE = 10;
114
118      virtual class test_base extends ovm_component;
119
120        typedef env #(top.DATA_SIZE, top.ADDR_SIZE) env_t;
121        env_t e;
122
123        function new(string name, ovm_component parent);
124          super.new(name, parent);
125        endfunction
126
127        function void build();
128          e = env_t::type_id::create("env",this);
129        endfunction
130
131      endclass
file: 07_complete_testbenches/03_tests/top.sv
```

Notice that the test is declared inside the top-level module, top. The testbench and components are best declared inside packages to facilitate reuse, but tests themselves typically rely on OOP inheritance for reuse. Having all the tests compiled in the top-level module provides the flexibility to choose which test to execute at run time, rather than requiring recompilation to switch between tests. It also allows the test to use parameter values specified in the module.

Default configuration parameters and other information can be defined in the base test or in the testbench. In this example, the test simply instantiates the testbench, which specifies default configuration for its children in the build() method:

```
73        function void build();
74
75            set_config_int("fpu_agent", "has_transport_master",
1);
76            set_config_int("fpu_agent", "has_monitor", 1);
77            set_config_int("fpu_agent", "has_talker", 1);
78            set_config_int("fpu_agent", "has_coverage", 1);
79
80            agent     = new("fpu_agent",  this);
81            reference = new("reference",  this);
82            c         = calc::type_id::create("calc", this);
83
84        endfunction
file: 07_complete_testbenches/03_tests/top.sv
```

Since configuration is hierarchical in OVM, a test may override a default configuration in the testbench, or it may set additional configuration. Typically, a test also sets factory overrides to swap new components into the environment to customize the behavior. In this example, the testbench configures the FPU agent to instantiate a coverage collector, but it is up to the test to specify which coverage collector to use. This is done in the build() method.

```
135       class test_one extends test_base;
136
137          `ovm_component_utils(test_one);
138
139          function new(string name, ovm_component parent);
140             super.new(name, parent);
141          endfunction
142
143          function void build();
144             super.build();
145             fpu_coverage::type_id::set_type_override(
146                                     fpu_ctrl_coverage::get_type());
147          endfunction
```

```
148
149      endclass
file: 07_complete_testbenches/03_tests/top.sv
```

Notice the calls to super.new() and super.build(). These ensure that the base test's underlying functionality is properly called, and they are required for every extension of base_test. The factory override specification tells the agent to instantiate the fpu_ctrl_coverage collector in place of its default fpu_coverage component. As shown in Section 7.2 above, this coverage collector ensures that all FPU operations are performed.

The same testbench may be used by a different test whose intent is to verify a different aspect of the FPU functionality. In this case, we may wish to use a different stimulus generator and check that interesting combinations of data values on the two operands are generated.

```
153      class test_two extends test_base;
154
155         `ovm_component_utils(test_two);
156
157         function new(string name, ovm_component parent);
158            super.new(name, parent);
159         endfunction
160
161         function void build();
162            super.build();
163            fpu_coverage::type_id::set_type_override(
164                                   fpu_data_coverage::get_type());
165            calc::type_id::set_type_override(calc2::get_type());
166         endfunction
167
168      endclass
file: 07_complete_testbenches/03_tests/top.sv
```

Notice that test_two is also an extension of test_base, but in this case, it instantiates both the new fpu_data_coverage collector and a new stimulus generator, calc2.

The test is actually executed via the run_test() task in the top-level module's initial block. This task uses the factory to create an instance of the test class specified either as a string argument or via the OVM_TESTNAME plusarg on the simulator command line. For this reason, all runnable tests must be registered with the factory. Once created by the factory, the test is run through its phases as any other OVM component, instantiating the testbench and executing the test. Additional tests may be added either as extensions to the base test or as further extensions to existing tests.

7.6   Summary

The OVM provides ways to specify analysis components to answer the does-it-work and are-we-done questions. Coverage collectors and scoreboards are created specifically to answer these questions by taking data produced by analysis ports and turning the data into useful information that guides the verification process. Protocol-specific questions may be answered by instantiating coverage collectors inside an agent, while additional application-specific functional questions may be answered by analysis components in either the testbench or the test.

Providing this clear separation between the structural testbench and the test facilitates reuse by allowing tests to modify the structure or behavior of components in the testbench. Having tests extend from each other or from a common base test greatly simplifies the task of creating additional incremental testcases to exercise different functionality. When sequences are added to the mix, as we shall see in the next chapter, the OVM provides ways for you to develop a vast array of tests without a lot of coding, making you more productive.

8

Sequences

To answer the are-we-done and does-it-work questions, we have to first stimulate the design in interesting ways. The ability of the testbench to exercise all the meaningful functions and corner cases of a design is dependent on the quality of the stimulus applied to it. Good quality stimulus is complete, yet spare. It causes the design to visit as many unique states as possible without undue repetition. Elements that generate good stimulus can be complex to build, so it is important to have a means for building stimulus generation elements in a modular, reusable fashion.

OVM provides a facility called *sequences* for building reusable stimulus generators. Sequences are objects that produce streams of *sequence items* for stimulating a driver. A sequence item is a transaction with some extra bookkeeping members.

8.1   Sequence Basics

A sequence bears a striking resemblance to a *functor*, although it's not exactly the same thing. In OOP lingo, a functor is an object that serves as a function replacement. SystemVerilog, which does not support operator overloading, does not allow you to create true functors; however, sequences come close. A sequence is an object, and like most OVM objects, it is derived from `ovm_object`. The essential feature of the `ovm_sequence_base` base class is that it contains a virtual task called *body()*. Executing a sequence means creating an instance of it and invoking its body task. The `body()` task is the reason for a sequence to exist.

Sequences are not components; therefore, they are not part of the component hierarchy. Sequences are associated with a *sequencer*, an object that is a component, and therefore, part of the component hierarchy. The sequencer

M. Glasser, *Open Verification Methodology Cookbook*, DOI: 10.1007/978-1-4419-0968-8_8,   165
© Mentor Graphics Corporation, 2009

provides a place to attach sequences and funnels sequence items to a driver. The sequencer also arbitrates among multiple sequences operating in parallel.

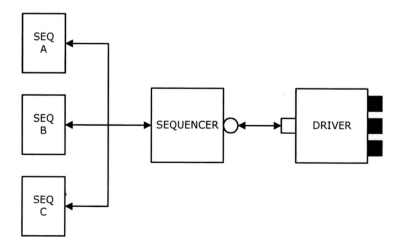

Figure 8-1 Relationship between Sequences, Sequencer, and Driver

The driver and sequencer talk to each other through a special TLM port called a seq_item_pull_port. In this, the most common use model, the driver pulls sequence items from the sequencer[1]. This port has an interface specifically designed for communication between a driver and a sequencer. The sequences associated with a particular sequencer also have a special API for communication. The API contains methods for requesting arbitration and granting access as well as sending and receiving sequence items. We'll look at these interfaces in more detail later in this chapter.

8.2   A Sequence Example

The most basic sequence configuration consists of three components: a driver, a sequencer, and a sequence. There are two APIs in play here: the sequence-sequencer API and the sequencer-driver API. To illustrate how sequences are constructed and initiated, let's consider the simple design in Figure 8-2.

---

1. An alternative use model has a push sequencer, which uses an ovm_blocking_put_port to put sequence items directly to the driver.

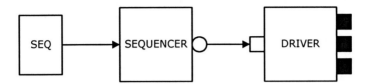

Figure 8-2 Simple Sequence Configuration

Because the sequencer and driver are components, they are instantiated and
connected in the top-level environment in the usual way.

```
129      function void connect();
130         drv.seq_item_port.connect(sqr.seq_item_export);
131      endfunction
file: 08_sequences/01_simple/top.sv
```

The sequence is instantiated when it is needed, in this case in the run() task.
The function start() serves to both associate a sequence with a sequencer
and initiate the execution of the sequence.

```
133      task run();
134         seq = new();
135         seq.start(sqr);
136         global_stop_request();
137      endtask
file: 08_sequences/01_simple/top.sv
```

start() is blocking; it returns only when the sequence completes. Now let's
look at the construction of the driver and the sequence. The sequence API
provides a means for requesting service, sending requests, and, as we will see
later, retrieving responses. Here is the sequence in our simple example:

```
92      class my_sequence extends ovm_sequence #(trans);
93
94      task body();
95
96         trans t;
97
98         for(int unsigned i = 0; i < 10; i++) begin
99            wait_for_grant();
100           t = new();
101           assert(t.randomize());
102           send_request(t);
103        end
```

```
104
105      endtask
106
107    endclass
file:  08_sequences/01_simple/top.sv
```

This sequence is derived from ovm_sequence and parameterized with the type of request transaction it will send to the driver. The meat of the sequence is in the body() task. This is where the work of the sequence is done. In this example, the main structure in body() is a loop that sends 10 transactions. To send a transaction, it first calls wait_for_grant() to request service. wait_for_grant() blocks until the sequence is ready to run. The sequencer arbitrates among multiple parallel sequences, and a call to wait_for_grant() puts an entry in the sequencer's arbitration queue and waits until the sequencer grants access.

Once access is granted, the sequence can then create the request sequence item and populate it with data. In our case, we simply randomize the item. send_request(), as the name implies, sends the request through the sequencer to the driver. send_request() is a nonblocking function, and it returns in the same delta cycle in which it is called.

To communicate with the sequencer, the driver has an ovm_seq_item_pull_port. This port is a special port, not one of the standard ports. It is constructed in the same manner as a standard TLM port, but it has a custom interface designed for communicating with a sequencer. The table below identifies the essential methods in the seq_item_pull interface.

task get_next_item(output T1 t)	Blocking function that retrieves the next request.
task try_next_item(output T1 t)	Pseudo-nonblocking function that retrieves the next request. Since it has to synchronize with the sequence process, it will consume at least one delta cycle.
function void item_done(T2 t = null)	Signifies to the sequencer that the item is complete.
function bit has_do_available()	Asks the sequencer if there is a sequence with a request pending.
function void put_response(T2 t)	Sends a response back to the sequence.

`task get(output T1 t)`	Retrieves the next request. This is equivalent to `get_next_item` (or `try_next_item`) followed by `item_done()`.
`task peek(output T1 t)`	Retrieves the next request without consuming it.
`task put(input T2 t)`	An alias for `put_response`. This function is in the interface to maintain consistency with the standard TLM interface functions.

Unlike a sequence, a driver is a component, and it has the usual constructor and phase callbacks. Since this is a simple example to illustrate the mechanics of sequences, the driver doesn't really drive signals on a bus. However, the the essential structure is the same as a real driver. The `run()` task is a forever loop that continually retrieves requests and simply processes them. In this case it just prints them.

```
59    class driver extends ovm_component;
60
61       ovm_seq_item_pull_port #(trans) seq_item_port;
62
63       function new(string name, ovm_component parent);
64          super.new(name, parent);
65       endfunction
66
67       function void build();
68          seq_item_port = new("seq_item_port", this);
69       endfunction
70
71       task run();
72          trans t;
73
74          forever begin
75             seq_item_port.get(t);
76             ovm_report_info("get request", t.do_sprint());
77             #1;
78          end
79
80       endtask
81
82    endclass
file: 08_sequences/01_simple/top.sv
```

There are several features of a driver intended to be driven by sequences that are different from a driver driven by a typical transaction-level model, as described in Section 5.1. First, the ingress port for transactions is a seq_item_pull_port. Second, the request and response types must be derived from ovm_sequence_item. Sequence items are much like transactions (that is, objects derived from ovm_transaction) except the ovm_sequence_item base class contains some additional members for use in routing the item through the sequencer to the driver and back.

Why must we use a special interface for connecting a driver to a sequencer rather than just use the standard TLM interfaces? The reason is that the seq_item_pull interface is designed to synchronize between two processes—the sequence process and the driver process.

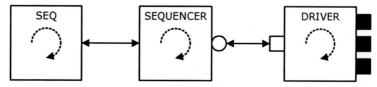

Figure 8-3 Processes Involved in Sequence Communication

Actually, there are three processes involved. Each of the objects involved in the communication—the sequence, the sequencer, and the driver—has a process, and those processes must all be synchronized in order for a sequence item to be transferred from the sequence to the driver. The sequence process is responsible for generating a sequence item. The sequencer process handles the arbitration of multiple parallel sequences, and the driver process is responsible for managing the bus.

Because there are multiple processes involved in moving a sequence item from the sequence to the driver, there is no true nonblocking method in the seq_item_pull interface. In the *get* example discussed in Section 3.4.2, both the producer and consumer are in the same process. In Section 3.5, we inserted a FIFO channel between the producer and consumer. The channel enables the producer and consumer to each have separate processes, and it synchronizes the two processes. That synchronization will consume at least one delta cycle. This is true even if both the producer and consumer use nonblocking methods to put things into the FIFO and get them out. In a practical sense, the sequencer operates as a channel between the sequence and the driver. Thus, the get(), get_next_item(), and try_next_item() methods in the seq_item_pull interface will always consume at least one delta cycle.

8.3   Anatomy of a Sequence

A sequence is conceptually a very simple object whose primary role is for its body task to generate a stream of sequence items, as we saw in the last section. However, it has a lot of parts that enable it to be used in a number of different ways. In this section we will discuss the various parts of the sequence object and sketch out its functionality beyond just execution of the body() task. The UML below shows how sequences and sequencers are organized.

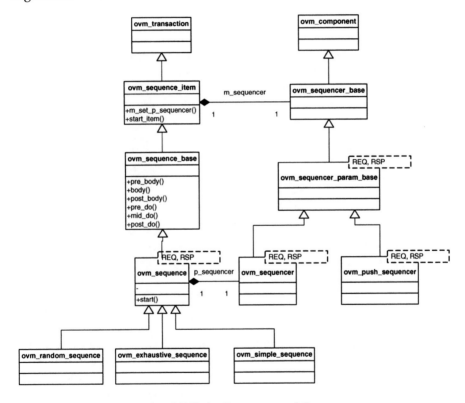

Figure 8-4 UML for Sequences and Sequencers

To review: *sequencers* are components; that is, ovm_sequencer is derived from ovm_component. The sequencer is parameterized with the request and response type of the items it will process.

Sequences, on the other hand, are not components. ovm_sequence is derived from ovm_sequence_item, which is in turn derived from ovm_transaction, which reflects the transient nature of sequences. Sequence items know which sequencer they are associated with via a special reference called m_sequencer. This enables sequence items and sequences to access the sequencer and use services it makes available.

A sequence may also have a reference called p_sequencer, which is also a reference to the sequencer. The difference between m_sequencer and p_sequencer is that m_sequencer is a reference to ovm_sequencer_base, the sequencer's base class, and p_sequencer is a reference to ovm_sequencer#(REQ,RSP), the derived parameterized sequencer class. The derived sequencer may have additional resources you add in.

The m_sequencer reference is set automatically when you call start(). Because the type of the p_sequencer is not known *a priori*, you must provide a declaration of p_sequencer and implement a function to set its value. In ovm_sequence_item, the base class for ovm_sequence, is the virtual function m_set_p_sequencer(). In your derived sequence, you will need to define the specific type of p_sequencer and provide an implementation of m_set_p_sequencer().

```
my_sqr_type p_sequencer;

function void m_set_p_sequencer();
   super.m_set_p_sequencer();
   assert($cast(p_sequencer, m_sequencer));
endfunction
```

As a convenience, the macro `ovm_declare_p_sequencer will define the type for you. You simply invoke it at the top of your sequence. Its argument is the type of the p_sequencer.

A sequence can run other sequences. In the example above, we showed the start() task being called from the run task of a component. The start() task can also be called from the body() task of another sequence. start() tasks can also be forked to allow multiple sequences to run in parallel. (We'll look at the details of how to manage parallelism amongst sequences in Section 8.6.) Running sequences in parallel and allowing them to call other sequences allows you to create arbitrary hierarchies of sequences. The sequence hierarchy, rooted at the sequencer, is not unlike the component hierarchy. Each sequence in the hierarchy has a location relative to the sequencer, and thus a unique path name. The function get_sequence_path() in ovm_sequence_item returns a string with the full path name of a sequence.

Since sequences are not components, and thus not anchored into the component hierarchy, they can be dynamic. Each sequence exists only as long as the body task continues to execute. When it terminates, so does the sequence. This is like the functor behavior described earlier. A functor comes into existence, executes its function, and then goes away to be garbage collected later. Of course, a sequence can be effectively static in that it comes into existence when the test starts, and goes away when the test concludes. The lifetime of a sequence is entirely dependent on its function.

When a sequence is started, pre_body() is called first, followed by body(), and then post_body(). pre_body() and post_body() are virtual tasks with default empty implementations. You can implement them as you please. Typically, pre_body() is used for one-time initialization and post_body() is for final clean-up.

8.4   Another Sequence API

In Section 8.2, we discussed how to use wait_for_grant() and send_request() to send sequence items from a sequence to a driver. While this is a perfectly valid way to transmit sequences, it's not entirely general. In this section we will discuss an alternate, more generalized way of sending sequences. The new API consists of three methods, create_item(), start_item(), and finish_item().

This alternate API is illustrated in an example that is a modified version of an earlier one shown in Section 7.4. Previously, we showed how to send transactions to a driver using a fixed stimulus generator: the random calculator. Here, we replace the fixed stimulus generator with a sequence.

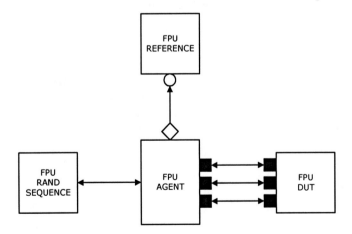

Figure 8-5 FPU Testbench with a Randomized Sequence

To make this replacement, the first thing we have to do is reconfigure the FPU agent to have a sequencer and not a master.

```
76          set_config_int("fpu_agent", "has_monitor", 1);
77          set_config_int("fpu_agent", "has_sequencer", 1);
78          set_config_int("fpu_agent", "has_talker", 1);
79          set_config_int("fpu_agent", "has_coverage", 1);
80
81
fpu_coverage::type_id::set_type_override(fpu_ctrl_coverage::get
_type());
file: 08_sequences/08_calc/top.sv
```

The has_sequencer flag instructs the agent to instantiate a sequencer. Then, in the run() task we use start() to initiate execution of the sequence.

```
98      task run();
99
100         fpu_seq_rand seq;
101
102         ovm_report_info("env", "start");
103
104         global_stop_request();
105         seq = new();
106         seq.start(agent.sequencer);
107
108         ovm_report_info("env", "finish");
109     endtask
file: 08_sequences/08_calc/top.sv
```

The sequence uses the alternate API to send a series of sequence items to the driver through the sequence. As you can see in the body() task below, we first create a new item, and then we communicate to the sequencer that we want to send an item to the driver by calling start_item(). To create an item, we pass its type handle to create_item(), which then uses the factory to instantiate a new instance. Embedded in start_item() is a call to wait_for_grant().

```
33      task body();
34
35          ieeeFloat f;
36
37          f = new();
38
39          for(int unsigned i = 0; i < 10000; i++) begin
40              assert($cast(req,
41                          create_item(fpu_request_item::get_type(),
42                              m_sequencer, "req")));
```

```
43          start_item(req);
44          assert(req.randomize());
45          finish_item(req);
46          get_response(rsp);
47        end
48
49      endtask
```

finish_item() calls send_request() and wait_for_item_done(). The contents of the sequence item are populated between the calls to start_item() and finish_item(), either through directed or randomized means. In the example above, we use both.

Why use this API instead of the previous one? First, recall that ovm_sequence is derived from ovm_sequence_item—a sequence *is* a sequence item. Also recall that a sequence can run another sequence. The start_item() API will look to see if the object passed in is a sequence item or a sequence. If it's a sequence, it will initiate execution on m_sequencer. If it's a sequence item, it will send it to the driver. So, start_item() provides a polymorphic way to initiate execution of sequences and sequence items.

8.5   Response Routing

Many bus protocols are bidirectional. In those protocols, masters send requests and receive responses. Slaves do the opposite. They receive requests and send responses. Drivers for bidirectional protocols have to return responses to a sequence. In the case where there are multiple sequences operating in parallel, there is the intuitive notion that each response sent back by the driver must be returned to the sequence that originated the request. The sequencer has some machinery for doing exactly that.

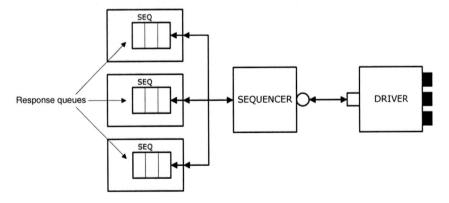

Figure 8-6 Sequences with Response Queues

Each sequence contains a response queue. The put_response() method, which is called by the driver, puts a response object in the queue for the sequence from which the request originated. The get_response() method in the sequence pulls the response object from its queue. The sequencer knows in which queue to put the response because the response item contains a sequence ID. The response item knows the sequence ID because the driver put the sequence ID into the response item. Unfortunately, this does not happen automatically. You have to make a call to move the sequence ID from the request into the response. Here's the body of a simple bidirectional driver:

```
137      task run();
138         request_item req;
139         response_item rsp;
140
141         forever begin
142            seq_item_port.get(req);
143            rsp = new();
144            rsp.copy_req(req);
145            rsp.set_id_info(req);
146            #1;
147            seq_item_port.put_response(rsp);
148         end
file: 08_sequences/02_bidir/top.sv
```

The call to set_id_info() copies sequence identification information from the request object into the response object and enables the put_response() method to know in which queue to put the response object.

The main code in the sequence is much like the code in the example earlier in this chapter. The exception is the get_response() call at the end of the loop.

```
169         for(int unsigned i = 0; i < 10; i++) begin
170
$cast(req,create_item(request_item::get_type(),m_sequencer,"req
"));
171            start_item(req);
172            assert(req.randomize());
173            m_sequencer.ovm_report_info("send",
174               req.do_sprint());
175            finish_item(req);
176            get_response(rsp);
177            m_sequencer.ovm_report_info("retrieve",
178               rsp.do_sprint());
179         end
file: 08_sequences/02_bidir/top.sv
```

The flow of control does a round trip from the sequence to the driver and back again. The activity is orchestrated by the sequencer. send_request() initiates data transfers from the sequence to the driver; put_response() transfers data and control back to the sequence. send_request() is nonblocking. Once it's called, data is sent through the sequence to the driver and the get (or get_next_item) task is allowed to return. get_response() is blocking, so it will wait until a response is available before it returns. This flow is illustrated in the diagram below:

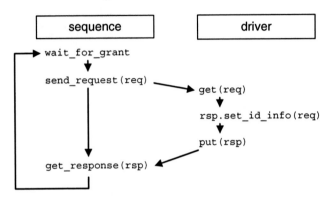

Figure 8-7 Flow of Control between Sequence and Driver

This lock-step relationship between sequences and drivers is intended for protocols where the request and response are synchronized, and there is one response for each request. OVM also provides an alternate control and data flow for situations where that might not be the case. Some protocols return responses for a group of requests instead of one response for each request. Some protocols return responses in a different order than the request stream. To handle these situations, the OVM sequence provides a response handler facility. Instead of calling get_response() to retrieve a response, you use a *response handler*. The response handler is a function that is invoked whenever a response from the driver becomes available.

Here's what the sequence looks like after it has been modified to use a response handler.

```
147    class my_sequence
148      extends ovm_sequence #(request_item, response_item);
149
150      int unsigned expected_responses;
151
152      function new(string name);
153        super.new(name);
154        expected_responses = 0;
```

```
155       endfunction
156
157       task pre_body();
158          use_response_handler(1);
159       endtask
160
161       task body();
162
163          request_item req;
164          response_item rsp;
165
166          for(int unsigned i = 0; i < 10; i++) begin
167            $cast(req, create_item(request_item::get_type(),
168                                   m_sequencer, "request"));
169            start_item(req);
170            assert(req.randomize());
171            req.seq_name = get_name();
172            m_sequencer.ovm_report_info("send",
req.do_sprint());
173              expected_responses++;
174              finish_item(req);
175          end
176
177          wait (expected_responses == 0);
178
179       endtask
180
181       function void response_handler(ovm_sequence_item
response);
182          response_item rsp;
183          assert($cast(rsp, response));
184          m_sequencer.ovm_report_info("retrieve",
185                                      rsp.do_sprint());
186          expected_responses--;
187       endfunction
188
189   endclass
file: 08_sequences/04_handler/top.sv
```

The first thing to notice is that the get_response call has been removed from
the main body loop. In its place, we've added a function called
response_handler. The response handler function is called by the sequencer
when the driver calls put_response() on its seq_item_port. The prototype
of ovm_sequence_base::response_handler() has ovm_sequence_item as
its argument. So, we must also use ovm_sequence_item in our
implementation and cast the argument to the response type. To let the
sequencer know that it should call the response handler when the driver
sends back a response, we turn it on using the call to
use_response_handler() in the pre_body() task.

The control flow is slightly different when we use a response handler. The response handler operates asynchronously from the sequence body. Since send_request() is nonblocking, the response handler will be invoked when the driver returns a response and the sequence is blocked. In our case, wait_for_grant() is the blocking call that allows the response handler to be invoked.

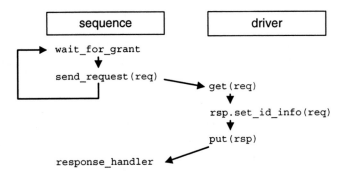

Figure 8-8 Flow of Control with a Response Handler

By using a response handler instead of calling get_response(), we have decoupled the collection of responses from the generation of requests. We can use the response handler to manage out-of-order responses or protocols that do not generate one response for every request.

8.6   Sequences in Parallel

Sequences can operate in parallel. Multiple sequences associated with the same sequencer can operate together to send a stream of sequence items into a driver. Running sequences in parallel is simply a matter of using fork to execute each sequence in separate, parallel processes.

```
147      task run();
148
149         seq1 = new("seq1");
150         seq2 = new("seq2");
151         seq3 = new("seq3");
152
153         fork
154            seq1.start(sqr);
155            seq2.start(sqr);
156            seq3.start(sqr);
157         join
158
159         global_stop_request();
160      endtask
```

```
file: 08_sequences/03_parallel/top.sv
```

In the run task above, three sequences are executed in parallel using the SystemVerilog `fork-join` construct. The `start()` method initiates execution of the sequence and associates it with a sequencer identified by the argument.

The sequencer arbitrates among all the executing sequences associated with it. When a sequence calls the blocking task `wait_for_grant()`, an entry is made in the sequencer's arbitration queue. The sequence chooses a sequence from the arbitration queue and allows it to execute by letting the task return in the selected sequence.

By default, the sequencer arbitrates the sequences in FIFO order. However, you have a lot of control over the specific order of parallel-sequence execution. The sequencer supports a variety of arbitration algorithms. They are summarized in the table below.

`SEQ_ARB_FIFO`	FIFO ordering. This is the default arbitration mode.
`SEQ_ARB_WEIGHTED`	Randomly choose the next sequence. Use weights specified by `wait_for_grant()` calls to bias the selection.
`SEQ_ARB_RANDOM`	Randomly choose the next sequence.
`SEQ_ARB_STRICT_FIFO`	All requests at the highest priority are granted in FIFO order.
`SEQ_ARB_STRICT_RANDOM`	All requests at the highest priority are granted in random order.
`SEQ_ARB_USER`	User supplies the arbitration algorithm.

To change the arbitration mode, call `set_arbitration()` on the sequence and specify the desired arbitration algorithm. A sequence specifies its weight when it requests arbitration by calling `wait_for_grant()`, which accepts an optional weight argument.

```
162        m_sequencer.set_arbitration(SEQ_ARB_WEIGHTED);
163        fork
164          A.start(m_sequencer, this, 100);
165          B.start(m_sequencer, this, 20);
166          C.start(m_sequencer, this, 10);
167        join
```

```
file: 08_sequences/05_arb/top.sv
```

The code fragment above uses the m_sequencer pointer in a sequence to call set_arbitration() and tell it which arbitration algorithm to use. Three sequences are spawned to run in parallel, each with a different weight. The weights are relative. The total weight for these three sequences is 100 + 20 + 10 = 130. The SEQ_ARB_WEIGHTED algorithm uses these weights to select the next sequence to run. Given these weights, sequence A will run 100/130 or 77 percent of the time, sequence B will run 20/130 or 15 percent of the time, and sequence C will run 10/130 or 8 percent of the time.

A sequence can specify that it doesn't want to give up access to the driver after only a single call to send_request(). It can lock the sequencer so that it will run continuously. A sequence can call lock() to retain ownership of the sequence and driver. lock() is a blocking call, and it will return when the request is granted. The sequence holds the lock until it calls unlock(). A slight variation of lock() and unlock() is grab() and ungrab(). These two functions work in the same way as lock() and unlock()—grab() is a blocking call that returns when the request is granted, and the lock is held until a corresponding call to ungrab() is made. The difference between the two is when the request is granted. Locks are arbitrated; they are put in the request queue along with other requests. When the lock comes to the head of the queue, then the sequencer grants the lock. Grabs, on the other hand, are not arbitrated. They jump right to the head of the queue. No sequence can have a higher priority than a grab.

By dividing the stimulus generation into small, modular sequences and executing them in parallel, you can build complex stimulus. Rather than building a single monolithic stimulus generator, you can more accurately mimic the environment that the DUT will run in.

8.7   Constructing APIs with Sequences

The essence of reuse, as we discussed previously, is to build components with well-defined interfaces. Because the interface is well defined, clients of the component know exactly what services the component provides and how to access those services. Dependencies are carefully contained. You can use the sequences facility to construct such interfaces, or APIs, for your test environment. You can create a library of sequences that enable access to the DUT or bus, and you can layer other libraries of sequences on top of that to build higher-level functionality.

We mentioned earlier that sequences had a passing similarity to functors. You can use that fact to construct APIs from a set of sequences. Consider that an API is a collection of methods. A sequence is an object whose role is to execute a single method, specifically its body() task. So, a set of sequences is essentially a set of methods. The protocol for initiating a sequence is a bit more verbose than just calling a task or function, but the idea is the same. The reasons for choosing a sequence-based API over a fixed component are based on all the sequence functionality we've discussed so far—sequences have facilities built in for managing concurrency; their lifetime is controlled by the nature of their functionality, not the containing component; they have a modular nature; and so forth.

As an example, we can construct a test API for the HFPB protocol by combining three sequences, one that performs a read operation, one that performs a write operation, and one that performs an idle. These are all the possible things that you can do with the HFPB protocol.

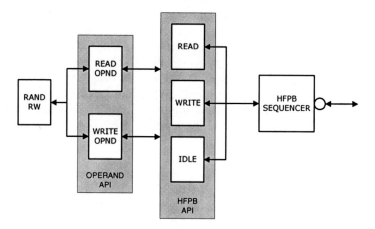

Figure 8-9 Using Sequences to Construct a Test API for HFPB

With those sequences in place, we can then layer an operand API that we will use for the FPU. An operand is one or more words. The API breaks operands into individual words that can be transmitted over an HFPB bus. On top of the operand API, we can build tests.

The design in Figure 8-9 uses a simple transaction-level memory to illustrate the construction of a sequence-based API and how it is formed into layers. In this example, we have a test sequence that randomly reads and writes operands. The operand reads and writes are implemented by the HFPB layer to form bus reads and writes. Each layer only needs to know about the layer below it. The top-level environment only needs to have knowledge of the test sequence. The test sequence only needs to know about the operand layer, and so forth. Each layer contains declarations and invocations of the sequences and the next, lower layer.

To start things off, the run() task in the top-level environment simply initiates the randomizing sequence, hfpb_seq_rand_rw.

```
77        task run();
78          hfpb_seq_rand_rw #(DATA_SIZE, ADDR_SIZE) rand_seq;
79          rand_seq = new("rand_rw");
80          rand_seq.start(sqr);
81          global_stop_request();
82        endtask
file: 08_sequences/06_api/top.sv
```

The body() task of the hfpb_seq_rand_rw starts with the declarations of the subordinate sequences it uses during its execution.

```
57        task body();
58
59          typedef hfpb_seq_read_operand
60             #(DATA_SIZE, ADDR_SIZE) read_seq_t;
61          typedef hfpb_seq_write_operand
62             #(DATA_SIZE, ADDR_SIZE) write_seq_t;
63
64          read_seq_t read_seq;
65          write_seq_t write_seq;
```

Note that the HFPB sequences are parameterized in the same fashion as the other HFPB components. The DATA_SIZE and ADDR_SIZE parameters are passed on to the operand sequences, which in turn pass them on to the read/write sequences, and then to the sequence items generated. The types of the parameterized sequence items then match the driver types.

The main part of the body() task uses the declared sequences to execute randomized reads and writes of operands.

```
68        for (int unsigned i = 0; i < iterations; i++) begin
69          case ($random & 1)
70
71            0 : begin
72                  assert($cast(read_seq,
73                      create_item(read_seq_t::get_type(),
74                      m_sequencer, "read_operand")));
75                  start_item(read_seq);
76                  assert(read_seq.randomize());
77                  finish_item(read_seq);
```

```
78                    end
79
80            1 : begin
81                    assert($cast(write_seq,
82                            create_item(write_seq_t::get_type(),
83                               m_sequencer, "write_operand")));
84                    start_item(write_seq);
85                    assert(write_seq.randomize());
86                    finish_item(write_seq);
87                 end
88          endcase
89        end
90
91      endtask
```

The `hfpb_seq_rand_rw` sequence uses the `start_item()`/`finish_item()` API for communicating with the sequencer. Similarly, the sequences in the lower layers use the same API to invoke their sequences.

8.8   Summary

Sequences provide a highly modular and flexible means for building complex stimulus generators. They provide sophisticated means for managing concurrency for handling responses. Using these functor-like objects, you can also build test APIs that encapsulate the low-level details of stimulus generation.

9

Block-to-System

Large-scale systems contain many elements of all different sorts—buses, bridges, processors, memories, special purpose slaves, and so forth. Each of these needs to be verified independently, and then all must be brought together and verified as a system. In this chapter, we explore block-level and system-level verification and how to share testbench components between them.

A single block can be any component of arbitrary complexity. It can be a simple adder or a complete DSP subsystem. The concept of a block is a design component that will become part of a larger system. When the block is integrated into a larger system, it's important to not lose the work done in building the block-level testbench. By reusing as many testbench elements as possible, you save the time of having to rewrite them. Also, results from the block-level tests can be verified again at the system level.

9.1   Reusing Block-Level Components

We will illustrate how to reuse a block-level testbench in a system-level testbench using an example, as we have done throughout this book. Previous chapters introduced a number of components for the FPU design and HFPB protocol. Figure 9-1 shows a system that's a bit more complex than we have seen so far, using both the HFPB protocol and the FPU.

The HFPB bus has two slaves, a transaction-level memory slave and a pin-level FPU slave. The FPU slave is really a bridge that connects the FPU protocol to the HFPB protocol. The FPU agent is used in monitor-mode. Although not shown in the diagram, the FPU agent has a coverage collector that is turned on. The FPU reference model takes transactions from the FPU bus and performs the same calculations as the DUT. It compares its results

M. Glasser, *Open Verification Methodology Cookbook*, DOI: 10.1007/978-1-4419-0968-8_9,

with the results from the bus to determine whether the RTL DUT performed the calculation correctly. Stimulus is generated entirely with layered sequences. The memory master and random calculator talk to the FPU-HFPB transport sequence, which in turn talks to the HFPB protocol layer.

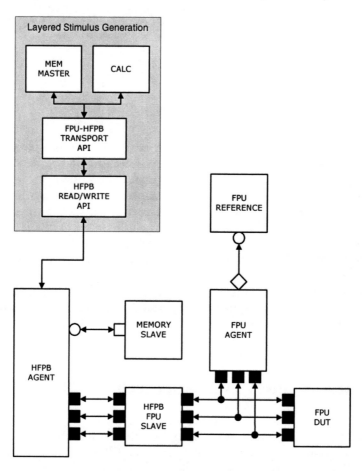

Figure 9-1 Complete System Testbench

In this design we have reused almost everything from previous block-level testbenches. The HFPB and FPU agents are the same as discussed previously. The memory slave, sequence-based test API, and reference model are also reused. The only thing new is the memory master and randomized calculator sequences. In Chapter 6, where we presented the FPU, we discussed a randomized calculator and a memory master as components, not as sequences.

Several things have enabled us to easily reuse components originally built for a block-level testbench in a complete system. The agent architecture is important here. It encapsulates everything about a particular protocol and provides a way to configure it for different situations. In the example shown in Figure 9-1, we have connected a transaction-level memory and a pin-level slave to the HFPB agent. Connecting them lets us reuse top-down transaction-level components and bottom-up RTL components in the same design. The agent supports stimulus through both a traditional transaction-level master or through sequences (but not both at the same time). Thus we can easily reuse stimulus built during the verification of component blocks.

9.2   Reusing Block-Level Testbenches

Even better than reusing components from a block-level testbench is reusing the entire testbench. Again, the architecture of agents and sequences lets us do just that. The agent's configurability lets us reuse the same component and make changes to suit a new application. Sequences, with their modular construction and support for concurrency, let us apply to the agent only the sequences required for a particular protocol.

We employ two essential concepts in integrating testbenches. First is to connect agents, possibly with a converter or adapter. The agents contain the communication protocol for each block, so it's logical that connecting agents is equivalent to connecting block-level DUTs. Each agent encapsulates knowledge about only one particular protocol, so it will be necessary to have a converter or adapter of some sort to facilitate communication between different protocols.

Second is to use sequences to generate stimulus. Because sequences can run concurrently, it's a simple matter of adding new sequences for the system-level tests; you can leave the block-level stimulus in place. There's no rewiring to be done.

In our example, we connect two testbenches: an HFPB memory testbench and an FPU testbench. Both of these are block-level testbenches, meaning they verify blocks that will be connected with a larger system. They don't make any assumptions about how they will be integrated into the system. Each testbench only knows about its own protocol and DUT.

Figure 9-2 shows the memory testbench. The memory is a transaction-level memory. Recall the internal topology of the HFPB agent from Figure 6-5. The sequencer sends sequence items to the driver, which converts them to pin-level protocol. The slave converts the pin protocol back to transactions, which are handled by the responder, which is the memory in this case.

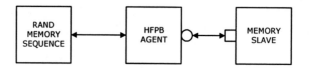

Figure 9-2 HFPB Memory Testbench

The FPU testbench shown in Figure 9-3 also uses sequences to send sequence items to the agent. The sequencer in the agent converts them to the pin-level protocol, which drives the DUT. The FPU reference model determines whether the DUT produces correct results. The sequence that drives this testbench generates random arithmetic calculations for the FPU to perform.

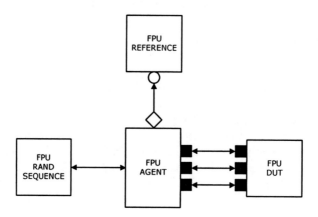

Figure 9-3 FPU Testbench

Now let's look at a design composed of both the memory and FPU. This design is an expression calculator. It takes as input ASCII strings containing infix algebraic expressions. The design parses the expressions and generates an intermediate program that runs on a processor. The processor executes the program to generate reads and writes on the bus in order to perform calculations and store and retrieve results from the memory. Figure 9-4 contains a block diagram of the complete system.

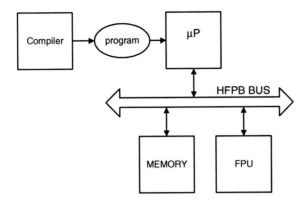

Figure 9-4 System Block Diagram

The compiler supports a mini-language for representing infix expressions, which includes assignments and a print statement. A typical infix expression might look like this:

```
14.2 * (19.0 + 3.2e3) / .002
```

The term *infix* refers to the fact that the operator appears between its two operands. The entire language can be described with the following BNF:

```
stmt         := print_stmt ; | assign_stmt ;
print_stmt   := print ID
assign_stmt  := ID = expr
expr         := term add_op term
add_op       := + | -
term         := factor mult_op factor
mult_op      := * | /
factor       := ( expr ) | - factor | NUM | ID
```

In this BNF, ID is an alphabetic identifier, and NUM is an integer or floating point constant. The compiler converts each ID to an address location, and when an identifier appears on the left-hand-side of an assignment, a new value is stored for it. And when it appears on the right-hand-side, its value is retrieved from memory. For example,

```
A = 92.4 * 3;
```

causes the result of 92.4 * 16 to be stored at the location identified symbolically by A. The following statement causes the value of A to be retrieved from memory and used in computing the expression.

```
B = 2 + (A / 7.14);
```

The result is then stored in the location identified by B. You can also print values using the print statement:

```
print A;

-> 277.2
```

Assignments, variable references, and print statements turn into memory reads and writes; numeric computations turn into FPU computations. They also turn into bus reads and writes, as the FPU is a slave on the HFPB bus, and transporting calculations and results to and from the FPU is done via bus operations.

To verify this design, we must build a testbench. We already have testbenches for the memory block and for the FPU block. We will illustrate how to integrate those with the complete system. The architecture for the system testbench is shown in Figure 9-5. The main bus is the HFPB bus, to which we must connect the FPU. The HFPB-FPU slave device serves as a bridge between the HFPB and FPU protocols. It converts HFPB requests to FPU request and FPU responses back to HFPB responses.

The FPU block does not know about the HFPB protocol, so the bridge is necessary. In practice, it is common for slaves to be built with a specific bus protocol in mind. In that case, the bridge is subsumed by the FPU block. Also in that case, the top-level environment does not have to specifically instantiate it. For the purposes of building examples to highlight OVM concepts, we chose to make it separate to emphasize the point that there must be an explicit connection between blocks.

The block testbenches still contain their sequences. These can be turned on or off. You may want to turn them on initially to make sure that the blocks still work as expected when connected to the system. Later, you can turn them off to operate the system-level tests without additional clutter. Should some anomaly appear, you can turn on the block-level sequences again to ferret out a bug.

In addition to the block-level sequences that come with the testbenches, we have new system-level sequences that exercise the entire system. These are instantiated in the top-level environment and connected to the appropriate agents. These sequences represent additional functionality that uses the connected elements; whereas, the block-level sequences assume only their own environment.

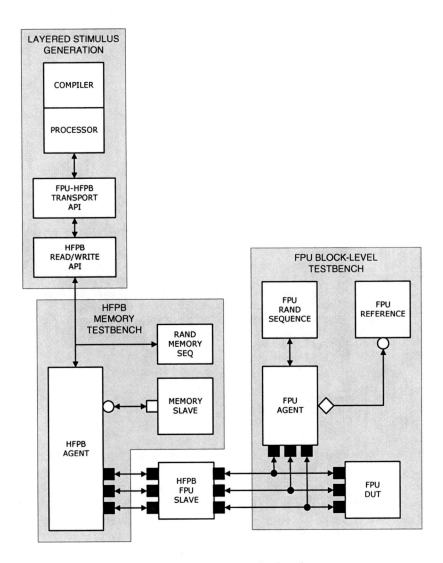

Figure 9-5 Connecting Testbenches

The code is surprisingly straightforward. There are only three components in
the top-level environment; the two subordinate testbenches and the HFPB-
FPU slave. Only the test sequence is invoked; it will invoke sequences in the
lower layers.

The FPU environment is the same as the one we saw in Chapter 7. It
instantiates the agent and the reference model, and it instantiates the

sequence. The only difference is, we've added a configuration item, run_sequences, which determines whether to run the block-level sequence. This configuration switch lets the system-level testbench turn on or off the block-level sequences. The run_sequences switch is also in the HFPB environment for the same reason.

The build() function in the top-level environment has a number of things in it. We'll walk through them.

```
193        function void build();
194
195           set_config_int("hfpb", "run_sequences", 1);
196           set_config_int("fpu", "run_sequences", 1);
197
198           set_config_int("*", fpu_slave_name, 1);
199
200           addr_map.add_range('h000, 'h00f, 1); // slave 1: fpu
201           addr_map.add_range('h010, 'hfff, 0); // slave 0: mem
202
203           set_config_object("*", "addr_map", addr_map, 0);
204           set_config_int("*", "rand_iterations", 500);
205           set_config_int("*", "mem_lower_bound", 'h80);
206
207           hfpb = new("hfpb", this);
208           fpu = new("fpu", this);
209           fpu_slave = new("fpu_slave", this);
210
211        endfunction
file: 09_block_to_system/02_expr/top.sv
```

First, we turn on the run_sequences switches for the subordinate environments. Next, we set the slave name for the FPU slave so that we can map the name to a slave identifier. After that, we set up the address map. This was moved from the HFPB environment to the top-level environment. The address map is a global resource, and it must be set from a vantage point where all the slaves are known. Next, we configure the randomized memory sequence with the number of iterations to run and the lower bound of the memory space it operates in. Finally, we instantiate the components.

The structure of the run() task should be familiar by now—we call global stop_request() followed by a sequence invocation. The only difference is the call to set_global_stop_timeout() before starting the sequences. The reason for that call is that subordinate environments rely on their coverage collectors to tell them when to stop. The global timeout instructs the stops to be released after the timeout expires, regardless of whether all the stop() tasks have returned.

```
213     task run();
214        string s;
215
216        ovm_report_info("env", "start");
217
218        seq = new();
219
220        set_global_stop_timeout(1ms);
221        global_stop_request();
222        seq.start(hfpb.agent.sequencer);
223
224        ovm_report_info("env", "finish");
225
226     endtask
file: 09_block_to_system/02_expr/top.sv
```

The global timeout is a safety valve that prevents the system from going into deadlock. If we turn off the block-level sequences, the system-level stimulus may not cover the blocks as expected. This is okay from a functional testing perspective, since the goal is to exercise the system, not specifically cover each block. However, the coverage information is still useful, so we leave coverage enabled. By setting the global stop timeout, we ensure that even if each block's coverage threshold is not reached, the system will terminate properly.

The top-level environment also participates in the global stop shutdown mechanism.

```
230     task stop(string ph_name);
231        seq.wait_for_sequence_state(FINISHED);
232     endtask
file: 09_block_to_system/02_expr/top.sv
```

The stop() task waits until the sequence reaches the FINISHED state. wait_for_sequence_state() is a blocking call that returns when the state named in the argument is reached.

9.3   Testing at the System Level

The type of stimulus that we generate for system-level testing is different from what we use for block-level testing. At the block level, we are concerned about only a single block. At the system level, we have to consider the entire system. Furthermore, the system is more than just a collection of blocks; it has additional functionality implemented using the smaller blocks. When designing tests for the complete system, it's reasonable to assume that each block has been tested thoroughly. After all, that's what the block-level testbench is for. You can assume that the necessary does-it-work and are-we-done questions were asked and answered satisfactorily.

So if each block is known to work correctly, then what's left to test when they are connected? Well, plenty. While the blocks themselves may have been tested thoroughly, the interaction between them has not. It's possible that a legal state in one block coincident with a legal state in another block may, in fact, constitute an illegal state. That is, a combination of states between each block may not be permitted.

Depending on the nature of a particular block, it may not be possible to exhaustively cover each and every state in a block. When designing the block-level stimulus and coverage model, you must draw lines about what is necessary to test and what is not. For example, in the FPU design, it's not possible to cover every possible value of the A and B operands. Each is 32 bits for a total of 64 bits of operands. To generate all the unique values for 64 bits would take considerable time. Further, there is the intuitive notion that it's not necessary to generate all those values. As a verification engineer, you have to figure out which values are significant and which are not. Later, when you integrate the block into a system, it's entirely possible that components in the system driving your block may pass it a value that was not tested at the block level. Along those same lines, the particular sequence of state transitions that the block takes may be somewhat different than what was exercised in the block-level tests. In theory, the block-level tests have exercised everything of significance. Yet it's still possible that a combination will occur in a system test that was not previously exercised.

It's important to retain the block-level coverage collectors to see if anything interesting, based on the coverage model, occurred in the system that did not occur in the block-level tests. Along the same lines, it's also important to keep the block-level scoreboards in place. These models determine whether the block functions correctly, which answers a does-it-work question. If they are reliable, then they can continue to monitor the operation of a block to flag any system-level failures. In the block-level testbenches in our example, the scoreboard and coverage collectors remain in place and are active.

Our small system is driven by a software element that compiles expressions. The stimulus we design for it must exercise the entire system and not necessarily focus on any particular block. Since the system is designed to parse and evaluate expressions, the stimulus problem becomes one of generating expressions and then evaluating them. To do that, we have a sequence that generates randomized ASCII strings, which represent valid expressions. Those expressions are passed through the chain of sequences until they become individual bus operations.

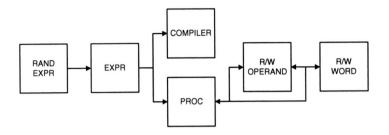

Figure 9-6 System-Level Chain of Sequences

The RAND EXPR sequence is the primary stimulus generator; it generates the randomized expressions. It passes them to EXPR, which takes the source string and passes it to the compiler, which parses it and converts it to a program and a symbol table. A data structure containing the program and data structure is passed to PROC, the processor, which executes the program. In so doing, it generates both operand reads and writes and word reads and writes. Operand reads and writes are converted to word reads and writes. This chain of sequences, built on top of the combined testbench infrastructure, represents our system functionality.

9.4   Summary

Integrating blocks into a system does not necessarily require rewriting testbench components. When you encapsulate protocol elements into a single agent and make sure that all components have well-defined interfaces, then you can easily connect block-level elements to form a system-level testbench. The block-level coverage collectors and scoreboards ensure that new, previously unforeseen stimulus is still counted and validated. Having reliable block-level testbench elements lets you focus on the system-level stimulus generation and functionality.

10

Coding Conventions

Good code doesn't happen by accident. It's the product of a good architecture and careful execution. Coding conventions can help improve the quality of your code by giving clear advice on how to structure your code and make sure the details are consistent.

10.1 Naming Scheme

Good quality code has a consistent look and feel. One of the ways to achieve a consistent look and feel is to use a consistent naming scheme. This section documents the naming scheme used for the examples in this book.

A name is constructed from three parts: the prefix, the main part, and the suffix. The main part of the name may consist of one or more words. All the parts—prefix, suffix, and words in the main part—are separated by underscores. Some sample names follow.

The following name has a main part of fifo and a prefix of tlm.

    tlm_fifo

The next one has all three parts, a prefix m_, a main part parent, and a suffix _p.

    m_parent_p

The following name has a main part with three words, but no suffix or prefix.

    finite_state_machine

M. Glasser, *Open Verification Methodology Cookbook*, DOI: 10.1007/978-1-4419-0968-8_10,   197
© Mentor Graphics Corporation, 2009

The next name also has no suffix or prefix, but the main part (top) consists of only one word.

```
top
```

The SystemVerilog OVM library is contained in a single package called ovm_pkg. The classes inside that package are prefixed with either ovm_ or tlm_. The reason for the tlm_ prefix is to match names in the OSCI TLM-1.0 standard which, of course, is rendered in SystemC.

When you construct names, avoid abbreviations, use complete words whenever possible, and distinguish names by case. This method is called the *general naming scheme*, and it forms the basis for the rest of the kind-specific naming schemes. A name should also identify the kind of thing it's naming. Below are rules for specific kinds of names:

- class names

  Class names use the general naming scheme. Classes that are part of a specific package or library should use the same prefix for all members of the package or library.

  ```
  ovm_analysis_port
  tlm_fifo
  ```

- local variables

  Local variables use the general naming scheme but have no prefix. They may have suffixes depending on the kind of object being named.

- integer indexes

  Use i, j, k for integer indexes. This is one place where single letter variable names are acceptable.

  ```
  int i;
  int j;
  for(i = 0; i < last; i++)
  {
      for(j = i; j < last; j++)
      matrix[i,j] = compute_entry(i,j);
  }
  ```

- class members

  Class members are another form of local variable. Instead of being local to a function or task, they are local to a class. To distinguish class members from local variables in a function, task, or method, use the local variable convention and add a prefix of m_.

```
class pc_bus_request
    addr_t m_address;
    data_t m_data;
    request_t m_request_type;
endclass : bus_request
```

- class methods

  For class methods, use a different prefix from the enclosing class, and group common methods with the same prefix.

```
class pc_bus_request
    addr_t m_address;
    data_t m_data;
    function set_addr(addr_t a);
    endfunction
    function set_data(addr_t a);
    endfunction
endclass : pc_bus_request
```

- local variables with suffixes

  It greatly improves the readability of a program if you are able to quickly understand something about the type or kind of object you are looking at in an expression without having to refer to the declaration.

- pointers

  Pointers appear in SystemC but not in SystemVerilog. Pointers use the local variable naming scheme and have a suffix of _p.

- handles

  Handles appear in SystemVerilog but not in SystemC. All instances of a class in SystemVerilog are referenced using class handles. Use the local variable naming scheme, and if no other suffix applies, add the _h suffix.

- type names

  For type names that are created with typedef, use the local variable convention and add the _t suffix.

```
typedef unsigned long int addr_t;
```

```
typedef sc_lv<16> bus_t;
typedef sc_port< sc_signal_in_if< sc_uint<32> > >
        bus_in_port_t;
typedef struct {bit [7:0] value} data_t;
```

- function/task/method names and formal arguments

  Functions, tasks, and methods (and their formal arguments) use
  the same convention as local variables — no prefixes or suffixes. A
  formal argument may be abbreviated if the abbreviation is
  derived from its type.

  ```
  function send(trasanction_t t, string parent);
  ```

- macros

  Macros are all uppercase letters, and words are separated by
  underscores. Distinguish between a macro and a named constant
  in SystemC. Macros are simply text to be substituted at the
  appropriate point in a program using a preprocessor. Named
  constants are constant values with a name known to the compiler
  and to the debugger.

  ```
  #define MAX_SIZE 100
  #define TRANSPORT(req,rsp) send(req);rsp=recv();
  ```

- parameters

  Parameters are all uppercase letters, and words are separated by
  underscores. An abbreviation may be used if it is derived from its
  type. In SystemVerilog, parameters or localparams are preferred
  over macros to reduce order-of-compilation issues.

  ```
  parameter type T = int;
  localparam MAX_SIZE = 100;
  ```

- enumeration types and enumeration members

  Enums need a suffix only if used as a defined type. In that case,
  use _e. For example:

  ```
  typedef {mode_unidir, mode_bidir, mode_off} mode_e;
  ```

  The members of enumerated types should have a common prefix
  that indicates their type.

  ```
  enum {color_red, color_blue, color_green} color;
  typedef enum {req_read, req_write, req_idle} req_e;
  ```

- interfaces
  - ◆ modports have the _mp suffix
  - ◆ interfaces have the _if suffix
  - ◆ virtual interfaces have the _vif suffix

```
interface bus_if;
...
endinterface : bus_if

virtual bus_if bus_vif;

class bus_if : public sc_interface
{
      161...
};
```

- packages

  Packages have the _pkg suffix. The main part of the package
  name should be the basis for the prefix of the names within the
  package.

- ports

  Transaction-level ports should use the same naming conventions
  as formal arguments to a function. Transaction-level ports should
  use the suffix _port or _export, as appropriate. Use _ap as a
  suffix for analysis ports. If you have only one analysis port in a
  module, which is quite common, just name it ap with no suffix or
  prefix.

```
sc_export<control_if> ctrl_export;
analysis_port error_ap, good_ap;
```

10.2   Global or Local?

In an object-oriented program, it's not always obvious in which scope to put
variables. For each variable, the answer to that question depends on the
lifetime of the variable and who will have access to it. In general, you want to
make variables that are in the inner-most scope possible and as local as
possible. The idea is that the more available you make variables, the more
likely some method, class, or module other than what you intend will modify
the variable, possibly with adverse consequences. This is a continuation of the
data-hiding concepts discussed in Chapter 2.

The remainder of this section offers guidelines to aid your decision-making
about where to put a declaration.

Class Interfaces. The interface to a class, the methods that enable external users to operate an object, should be public. Everything else should be local (private). Variables should all be private. If it's necessary to access them externally, this should be done with *accessor functions*, functions whose entire role is to provide read or write access to a class member. Consider as an example the following code:

```
class some_class;

    local int a;

    function void set_a(int a_arg);
        if(a_arg & 1) begin
            $display("a must be even");
            return;
        end
    a = a_arg;
    endfunction

    function int get_a();
        return a;
    endfunction

    ...
    endclass
```

The class some_class has two accessor functions, set_a() and get_a(). These functions guarantee that the variable a is accessed in only the allowed manner and no other way. set_a() guarantees that a is always set to an even value. It's not possible to set a to a value that's not even.

There are some exceptions to the rule that all variables should be private. One exception is for component ports and exports. These must be public so that external components can connect to them. The internal variables of these objects are private, so we still have preserved data-hiding. Another exception is for transactions. This exception is more subjective, and application of accessor functions should still be considered. The following transaction has a number of internal fields.

```
class trans extends ovm_transaction;
    bit [7:0] data
    bit [15:0] addr;
    int target;
    operator_e op;
    status_e status;
endclass
```

Since the purpose of the transaction object is to deliver data from one component to another, it can be a bit overly verbose and somewhat inconvenient to create a set_*() function and a get_*() function for each member. Providing both a set_target(), which assigns a value to target, and get_target(), which retrieves a value from target, for example, is semantically equivalent to just using assignment statements to modify or retrieve the value of target. In that case, the accessor functions provide no value. Consider a different example:

```
class packet extends ovm_transaction;
    int target;
    local bit [7:0] payload [255];
    local bit [15:0] error_correction_code;

    function void set_payload(bit [7:0] p [255]);
        payload = p;
        compute_error_correction_code();
    endfunction

    function bit [15:0] get_error_correction_code();
        return error_correction_code;
    endfunction
endclass
```

The class packet has a set_payload() accessor function, which not only sets the value of payload, it also computes the error correction code associated with the payload. The class does not provide a way to set the value of error_correction_code, which is local, except by setting the payload and computing a new error correction code. The set_payload() accessor is responsible not only for assigning a variable, but also for computing another (local) member in the class.

Loop Variables. It's best to keep loop variables in the function in which they are used, even in the loop scope. SystemVerilog provides a way to declare a loop variable at the top of the loop in which it is used. You should use this when you can.

```
for (int i; i < 100; i++) begin
    ...
end
```

Here, the loop variable is declared in the inner-most scope possible. If you have several loops in the same function, it's better to declare the loop variables once and not have to repeat the declarations. Besides saving a bit of typing, it guarantees that the type is consistent.

```
int unsigned i;

for(i = 0; i < max; i++) begin
    ...
end

for(i = 0; i < max; i++) begin
    ...
end
```

In any case, there's no reason to declare loop variables as class members. They should be in the function in which they are used. Even if you have multiple functions with loops, declare the loop variables in the functions. This removes the implication that the loop variables have anything to do with the other members in the class. It also prevents sharing problems in the event tasks with processes try to share the same loop variable.

Global Variables. Should you ever use global variables? Well, usually not. If you find yourself making a variable global, think about it carefully before you commit. The main problem with global variables is that they are not threadsafe. That is, two threads can be updating a global variable, unbeknownst to each other. Consider two threads, A and B, each of which maintains a global variable called X. A computes a value for X and assigns it to X accordingly. B now computes a new value and assigns it to X. Since A *believes* X to be a different value, it can make some assumptions, which, because X has changed values, are now wrong.

There are times when global variables make sense. In OVM, we have the global report server, for example. It's important for this variable to be global because it must be accessible everywhere. We prevent multiple copies of the report server from being created by making it a singleton object. We can get away with it being global because the report server has only functions, no tasks, so there is no possibility of any operation on the report server blocking or consuming time. So, all operations can be considered atomic. All access to the report server is through those functions, so there is no opportunity for multiple clients to create a conflict by setting variables in unexpected ways.

10.3   Objects

The OVM base class library defines a collection of objects of various types. This section presents some recommendations for coding those objects in a consistent manner.

10.3.1 Components

Components, objects derived from `ovm_component`, are one of the essential elements of an OVM testbench. Generally, your testbench will be constructed as a collection of interconnected components. So, it's important to organize components consistently. Within components, the primary organization is the set of phase callback functions. Schematically, it is best to organize components by putting items in a well-defined order.

1. declaration macro(s)
2. external interfaces
3. internal channels
4. configuration items—variables whose values are obtained through the configuration system
5. local variables
6. constructor
7. phase callbacks
8. local methods

Here's an example:

```
class my_component extends ovm_component;

    `ovm_component_utils

    // external interfaces
    ovm_get_port#(trans) get_port;

    // internal channels
    tlm_fifo#(trans) fifo;

    // configuration items
    int unsigned size;

    // local variables
    int i, j;

    // constructor
    function new(string name, ovm_component parent);
        super.new(name, parent);
    endfunction

    // phase callbacks
    function void build();
    endfunction

    function void connect();
    endfunction
```

```
function void end_of_elaboration();
endfunction

function void start_of_simulation();
endfunction

task run();
endtask

function void extract();
endfunction

function void check();
endfunction

function void report();
endfunction

// local methods
task send_to_bus();
endtask

endclass
```

The macro `ovm_component_utils generates some boilerplate code that is useful in all components. It creates the following code:

- Factory registration
- get_type() function
- get_type_name() function

For parameterized components, use `ovm_component_param_utils instead. The difference is that `ovm_component_utils provides the get_type_name() function and registers the component with the string-based factory as well as the type-based factory. The `ovm_component_param_utils assumes that it's not possible to create a unique string for a parameterized class, so it doesn't try. Instead, it just registers the component with the type-based factory and creates the get_type() function, but not the get_type_name() function.

Components have a standard constructor with a name and parent argument. Resist the temptation to add parameters to the constructor. Instead, use the configuration facility to pass data into a component. That way, you won't create an unnecessary dependency between a component and where it is instantiated. Furthermore, the factory is set up to create components using name and parent arguments, it currently does not support arbitrary arguments.

The `build()` function is a good place to retrieve configuration items. Since build is a top-down phase, it's also a good place to set configuration information to be passed to subordinate components. Construct your build function with `get_config_*` calls first, followed by instantiations of subordinate components, followed by `set_config_*` calls. Retrieving configuration items before instantiating subordinate components allows you to use the configuration information when constructing those components. Thus, you can create configurable topologies.

10.3.2 Sequences

Sequences are another kind of object that will contain a lot of your testbench code. Just like with components, organizing the code inside the class in a consistent manner will help you and others reading the code to find things and understand the structure easily. Code the items in a sequence in the following order:

1. `'ovm_sequence_utils` macro
2. declaration for child sequences
3. local variables
4. `pre_body()` task
5. `body()` task
6. `post_body()` task
7. response handler
8. local methods

10.3.3 Transactions and Sequence Items

Components are persistent structural objects; that is, they are created at the beginning of simulation and persist until the end. Sequences are semi-persistent; they remain in place until the `body()` task completes. This could be long enough to send a single sequence item or for the duration of the entire simulation. Transactions and sequence items are transient. They carry information between components and are released to be garbage collected once they deliver their payload. During their trip through the system, these objects can be copied, cloned, compared, or printed. So, they need methods to perform these functions. The ingredients that go in sequence items and transactions, in order, are:

1. `'ovm_object_utils` or `'ovm_object_param_utils` macro
2. `this_type` typedef
3. copy function
4. clone function
5. print function

6. compare function

The `this_type` typedef is useful for parameterized classes. You can use `this_type` in all the places where the class type is needed, and doing so can improve the clarity of the code in the case where there are multiple parameters. For example:

```
class trans#(type T=int, type R=int, int unsigned I=0)
   extends ovm_transaction;

   typedef trans#(T, R, I) this_type;

   ...

endclass
```

Here's what a prototypical transaction looks like:

```
class transaction#(type T=int) extends ovm_transaction;

   typedef transaction#(T) this_type;

   // declare transaction members here

   function ovm_object clone();
      this_type t = new;
      t.copy(this);
      return t;
   endfunction

   function void copy(input this_type t);
   endfunction

   function bit comp(input this_type t);
   endfunction

   function string sprint();
      string s;
      $sformat(s, ...);
      return s;
   endfunction

endclass
```

10.4   Packages

Packages in SystemVerilog provide a means for creating distinct namespaces. This is a powerful tool for managing large bodies of code. You can collect groups of related classes and types and make them available as single entities.

Using packages, you can maintain separation between these groups. You can hide objects and types that are needed for the implementation of the visible members but themselves are not visible. Use packages liberally to prevent unintentional interactions among objects and to simplify things for those who use your testbench components.

The package name itself is in the global namespace, but the items contained inside the package are in their own namespace. To make an item (or symbol) visible outside the package, you must import it. For example, to import an item called `driver` from `my_package`, use the following statement.

```
import my_package::driver;
```

To import all the symbols in `my_package`, use an asterisk (*).

```
import my_package::*;
```

This approach provides considerable control over the visibility of symbols and makes it obvious in programs that use packages whose symbols are "in play."

Even though the package forms a namespace, prefix all the symbols that will be visible externally (that is, those that users can import) with a common prefix. For example, all of the symbols in the OVM package have the prefix `ovm_` (with a few minor exceptions).

All of the components for a protocol should logically be grouped together, and you can use a package to form this grouping. The package can also contain any types or objects that are shared among the components in the package and which may not be externally visible. A good way to achieve this objective is to put each component in a single file and to `include those files in a package shell.

```
package abc_pkg;
    `include "abc_common.svh"
    `include "abc_driver.svh"
    `include "abc_monitor.svh"
    `include "abc_agent.svh"
endpackage
```

The package shell is put into its own file. The file `abc_common.svh` contains code that is common among the other components in the packages but might not be used directly by the user of the package.

10.5   Comments

Comments are the most subjective of coding style elements. Most programmers have their own feelings about what makes a good comment style. However you use comments, they should enhance the clarity and readability of the code and not obfuscate it. Here are some simple rules for making comments enhance your code and not detract from it.

- Don't replicate code in your comments. In other words, don't state the obvious.

  The comment here is unnecessary:

```
// Add one to counter
counter = counter + 1
```

- Put a comment header on each class and on major functions, particularly interface methods in classes.

  This is a convenient place to summarize the role of the class or method and an obvious place to look when reading code to find out what the class or method does.

```
//----------------------------------------------------------------
// env
//
// Top-level environment. Contains the bus agent and
// instantiates the test.
//----------------------------------------------------------------
class env extends ovm_component;
    ...
endclass
```

- Explain passages that implement complex or non-obvious algorithms.

  When you look at code that you have written and you find yourself having to spend a few minutes to reconstruct your thinking in order to understand what a particular passage is doing, then it is likely that this piece of code warrants a comment.

10.6   Summary

A good coding style that includes conventions for data and file names; use of globals, constants, statics, ports and exports; and all the other ingredients discussed in this chapter contributes to making the code accessible and can greatly aid the integration of code written by different programmers. When multiple engineers are working together on the same project, agreed-upon conventions will save considerable time and stress when it's time to combine the ingredients in a single recipe (system).

Just like any writing, when you apply a consistent style to your code, you improve the ability for someone reading it to understand it quickly and accurately. That person might be you!

Afterword

Verifying a complex system is a non-trivial problem. It requires a deep understanding of software engineering and electrical engineering, as well as extensive knowledge of the system being verified and the protocols it uses to communicate with the external world. Additionally, it requires some creativity and guile to design tests that effectively exercise the DUT to prove that it works correctly.

This text has presented the essential elements of the OVM—components, transaction-level interfaces, ports and exports, sequences, and so forth—that you can use to design and construct testbenches. The OVM is not just a library of parts; it is also a methodology for approaching complex verification problems. However, the methodology is just a starting point. There is a lot of room for creative application of the OVM. Every design house and design team has its own styles and conventions for building systems, and every design, even those derived from other designs, has its own unique verification challenges. So testbench architectures will vary widely based on the nature of the design, the verification requirements, the culture and training of the verification team, and the history of the project.

The methods and techniques expressed in this text are not dogma; rather they are a conceptual framework for experimentation and development of new methods and techniques. I encourage readers to explore new ways to apply the OVM to your own verification problems. Further, I encourage readers to exchange ideas on *www.ovmworld.org*. Open discussion and exchange will help to advance the OVM, which in turn will help to improve verification practice.

The Verification Methodology team at Mentor Graphics closely monitors the discussion on *www.ovmworld.org* and will respond to questions and bug reports and will participate in discussions there. If you wish to communicate directly with the team, you can do so by sending E-mail to *OVM_Cookbook@mentor.com*.

A

Graphic Notation

Throughout this book we illustrate examples with diagrams that show verification components and their interconnections. We use a schematic-like notation for these diagrams that combines both data flow and control flow concepts.

Traditional RTL schematic notation is *data-flow* oriented. Components have pins connected by nets. Pins have direction—they can be inputs, outputs, or bidirectional—and they must be connected to other compatible pins. For example, an output of one component must be connected to an input of another component. In systems that have transaction-level components, we need to describe *control flow* as well as data flow. Transaction-level models are constructed of function calls. Activity generated as functions in one component will call functions in other components. Control flow refers to who calls whom.

Connecting separate components through well-defined interfaces is a key tenet of the OVM, and those ideas are reflected in our notation. The graphical notation has three parts: components, interfaces, and interconnect.

A.1   Components

A component is represented using a box.

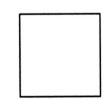

Figure A-1 Component Symbol

Components are objects such as modules, interfaces, program blocks, or classes that can be instantiated. Components often have free running threads. Sometimes, knowing the location of threads in a design or testbench is important to understanding the design. To show a component that has one or more threads, we use a circular arrow.

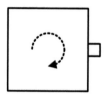

Figure A-2 A Component with a Thread

A.2   Interfaces

Interfaces are the externally visible connections to components. All of a component's behavior is accessible and visible only through its interfaces. First, is the familiar pin interface.

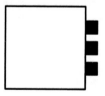

Figure A-3 A Component with a Pin Interface

The small black boxes on the right side of the component represent pins.

Whereas pin interfaces move data represented at the bit level between components, transaction interfaces move high-level data between components.

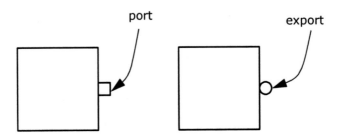

Figure A-4 Transaction-Level Interfaces

Figure A-4 represents two variations of transaction interfaces: a port and an export. The component on the left has a transaction port and the component on the right has an export. An export represents the *provides* sides of an interface and a port represents the *requires* side. A good way to think about transaction ports is as a set of unresolved function calls that are resolved by exports. Ports and exports are complements of each other; ports connect to exports. You cannot connect an export to an export or a port to a port.

The port/export notation identifies the flow of control between components. Since a port interface calls functions on an export, flow of control moves from ports to exports.

A.3   Interconnect

Just like with traditional schematics, we use lines between interfaces to show the interconnection amongst components. The addition of arrow heads allows us to represent data flow.

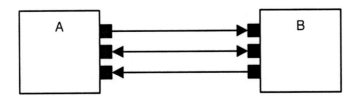

Figure A-5 Pin-Level Data Flow

Arrows between pins show the direction data flows between components. The figure above shows, from top to bottom, flow from A to B, bidirectional flow between A and B, and flow from B to A.

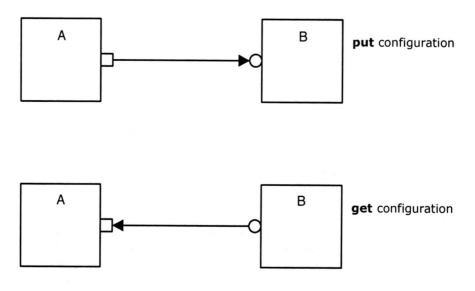

Figure A-6 Transaction Data Flow

Figure A-6 illustrates two configurations, each with the same transaction interfaces, but with different data flow. In both configurations, a function in B is invoked by A; that is, A initiates activity in B. A is the *initiator* and B is the *target*. In the top configuration, A moves data to B. This is called a *put* operation. In the bottom configuration, A moves data from B back to itself. This is called a *get* operation.

A.4   Channels

Transaction-level components often communicate through channels. A channel is a component that defines the semantics of the communication. One of the most common channels used is a FIFO. FIFOs are used to throttle communication between two transaction-level components. To show this in a netlist, we show a small box between components to represent the FIFO.

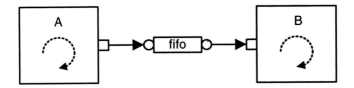

Figure A-7 Two Components Communicating through a FIFO

A FIFO, as with other communication channels, exports an interface. However, in the interest of keeping the diagram uncluttered, the circles on the channel exports are optional and often omitted. Just like vowels in Hebrew, exported interfaces on channels are obvious to conversant readers.

Figure A-7 shows two components, each with its own thread, and each with a transaction port that connects to an intervening channel. Component A puts transactions into the FIFO channel, and component B gets transactions from the same channel. The data flow arrows, in addition to the transaction ports, tell us which components are doing gets and which are doing puts. A has a thread, a transaction port (as opposed to an export), and an arrow leading away from it. That tells us that A is putting transactions into the channel. B also has a thread and a transaction port, but the data flow arrow is leading into the component instead of away from it. That tells us that B is getting transactions from the channel.

A.5   Analysis Ports

Analysis ports are a kind of transaction-level port used for communicating information between components involved in the operation of the DUT and components used to analyze activity. The symbol for an analysis port is a diamond. Analysis ports are connected to a component with an analysis interface. This could be an analysis FIFO or a component with an analysis interface.

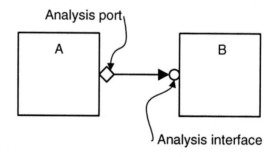

Figure A-8 Analysis Port Connected to an Analysis Interface

A.6   Summary

The OVM graphic notation is an extension to traditional RTL schematic notation. The extensions let us show transaction-level components, such as initiators, targets, interfaces, and channels, along with control and data flow between components. Using this notation, we can combine transaction-level and RTL components on the same diagram, which is important for diagramming testbenches.

Bibliography

Standards

1. IEEE standard 1800-2005, *IEEE Standard for SystemVerilog Unified Hardware Design, Specification, and Verification Language*, November 2005.
2. IEEE, standard 1666-2005, *IEEE Standard SystemC Language Reference Manual*, March 2006.
3. Open SystemC Initiative, *OSCI TLM-1.0 Transaction Level Modeling Standard*. SystemC kit with white paper available on *http://www.systemc.org*.

Functional Verification

4. Janick Bergeron, *Writing Testbenches: Functional Verification of HDL Models*, Second edition, Kluwer Academic Publishers, 2003.
5. Andreas S. Meyer, *Principles of Functional Verification*, Elsevier Science, 2004.
6. Harry D. Foster, Adam C. Krolnik, David J. Lacey, *Assertion-Based Design*, 2nd Edition, Kluwer Academic Publishers, 2004.
7. Chris Spear, *SystemVerilog for Verification: A Guide to Learning the Testbench Language Features*, Springer 2006

SystemC

8. Thorsten Grotker, Stan Liao, Grant Martin, Stuart Swan, *System Design with SystemC*, Kluwer Academic Publishers, 2002.
9. David C. Black and Jack Donovan, *SystemC: From the Ground Up*, Kluwer Academic Publishers, 2004.
10. J. Bhasker, *A SystemC Primer*, Star Galaxy Publishing, 2002.
11. Frank Ghenassia (ed.), *Transaction-Level Modeling with SystemC: TLM Concepts and Applications for Embedded Systems*, Springer, 2005.

C++ and Object-Oriented Programming

12. Stanley B. Lippman, *Inside the C++ Object Model*, Addison-Wesley, 1996
13. Bjarne Stroustrup, *The C++ Programming Language*, Third Edition, Addison-Wesley, 1997.
14. Gregory Satir, Doug Brown, *C++: The Core Language*, O'Reilly & Associates, Inc., 1995.

15. Erich Gamma, Richard Helm, Ralph Johnson, John Vlissides, *Design Patterns: Elements of Reusable Object-Oriented Software*, Addison-Wesley, 1995.
16. Andrei Alexandrescu, *Modern C++ Design: Generic Programming and Design Patterns Applied*, Addison-Wesley, 2001.
17. Martin Fowler, *UML Distilled: a brief guide to the Standard Object Modeling Language, Third Edition*, Addison-Wesley, 2004

Programming Style

18. Steve McConnell, *Code Complete, Second Edition*, Microsoft Press, 2004.
19. Herb Sutter, Andrei Alexandrescu, *C++ Coding Standards: 101 Rules, Guidelines, and Best Practices*, Addison-Wesley, 2005.

Miscellaneous

20. Niklaus Wirth, *Algorithms + Data Structures = Programs*, Prentice-Hall, Inc., Englewood Cliffs, New Jersey, 1976

Index

B

C

N

O

LaVergne, TN USA
15 October 2009
161044LV00002B/1/P